Lang Kurt

Advances in Geophysical and Environmental
Mechanics and Mathematics

Series Editor: Professor Kolumban Hutter

Christian Kharif · Efim Pelinovsky · Alexey Slunyaev

Rogue Waves in the Ocean

Prof. Christian Kharif
IRPHE
Technopole de Chateau-Gombert
49 rue F. Joliot Curie
13384 Marseille
BP 146
France
kharif@irphe.univ-mrs.fr

Prof. Efim Pelinovsky
Russian Academy of Sciences
Inst. Appl. Physics
Ul'yanov str. 46
Nizhny Novgorod
Russia 603950
Pelinovsky@hydro.appl.sci-nnov.ru

Dr. Alexey Slunyaev
Russian Academy of Sciences
Inst. Appl. Physics
Ul'yanov str. 46
Nizhny Novgorod
Russia 603950
Slunyaev@hydro.appl.sci-nnov.ru

ISBN: 978-3-540-88418-7   e-ISBN: 978-3-540-88419-4

Advances in Geophysical and Environmental Mechanics and Mathematics

ISSN: 1866-8348   e-ISSN: 1866-8356

Library of Congress Control Number: 2008936876

© Springer-Verlag Berlin Heidelberg 2009

This work is subject to copyright. All rights are reserved, whether the whole or part of the material is concerned, specifically the rights of translation, reprinting, reuse of illustrations, recitation, broadcasting, reproduction on microfilm or in any other way, and storage in data banks. Duplication of this publication or parts thereof is permitted only under the provisions of the German Copyright Law of September 9, 1965, in its current version, and permission for use must always be obtained from Springer. Violations are liable to prosecution under the German Copyright Law.

The use of general descriptive names, registered names, trademarks, etc. in this publication does not imply, even in the absence of a specific statement, that such names are exempt from the relevant protective laws and regulations and therefore free for general use.

Cover Design: deblik, Berlin

Printed on acid-free paper

9 8 7 6 5 4 3 2 1

springer.com

Preface

"It came from nowhere, snapping giant ships in two. No one believed the survivors... until now"
—New Scientist magazine cover, June 30, 2001

Rogue waves are the focus of this book. They are among the waves naturally observed by people on the sea surface that represent an inseparable feature of the Ocean. Rogue waves appear from nowhere, cause danger, and disappear at once. They may occur on the surface of a relatively calm sea and not reach very high amplitudes, but still be fatal for ships and crew due to their unexpectedness and abnormal features. Seamen are known to be unsurpassed authors of exciting and horrifying stories about the sea and sea waves. This could explain why, despite the increasing number of documented cases, that sailors' observations of "walls of water" have been considered fictitious for a while.

These stories are now addressed again due to the amount of doubtless evidence of the existence of the phenomenon, but still without sufficient information to enable interested researchers and engineers to completely understand it. The billows appear suddenly, exceeding the surrounding waves by two times their size and more, and obtaining many names: abnormal, exceptional, extreme, giant, huge, sudden, episodic, freak, monster, rogue, vicious, killer, mad- or rabid-dog waves, cape rollers, holes in the sea, walls of water, three sisters, etc. Freak monsters, though living only for seconds, were able to arouse the superstitious fear of the crew and cause damage to the ship and death to heedless sailors. All these epithets are full of human fear and frailty.

Serious studies of the phenomenon started about 20–30 years ago and have intensified during the recent decade. The research is being conducted in different fields: physics (search of physical mechanisms and adequate models of wave enhancement and statistics), geoscience (determining the regions and weather conditions when rogue waves are most probable), and ocean and coastal engineering (estimations of the wave loads on fixed and drifting floating structures). Thus, scientists and engineers specializing in different subject areas are involved in the solution of the problem. Freak waves annually become the subject of special sessions at the European Geophysical Union Assembly (2001–2008); Ifremer (France) organized workshops "Rogue Waves" in Brest (2000, 2004, 2008) 'Aha Huliko' (a Hawaiian Winter

Workshop in 2005) and a workshop held the same year by the International Centre for Mathematical Sciences (Edinburgh) were also dedicated to this phenomenon.

We start this book with a brief introduction to the problem of freak waves, aiming at formulating what is understood as *rogue* or *freak* waves, what consequences their existence imply in our life, and why people are so worried about them.

Chapter 1 is devoted to observations and measurements of freak waves. After some citations of personal descriptions of unexpectedly high waves, we proceed to speak about available instrumental measurements of rogue waves that can allow some quantitative analysis. In spite of recent success in developing the measuring systems, there are difficulties and problems that embarrass the high wave registration and analysis; they will be also discussed in Chap. 1.

Two approaches to the rogue wave description (deterministic and statistical) are discussed in Chap. 2, where some definitions and a mathematical toolkit are provided that are necessary for the following chapters. A brief survey of the physical mechanisms that have been already suggested as possible explanations of the freak wave phenomenon completes Chap. 2. They are:

- wave-current interaction
- geometrical (spatial) focusing
- focusing due to dispersion (spatio-temporal focusing)
- focusing due to modulational instability
- soliton collision
- atmosphetic action

This brief survey anticipates the detailed description given in Chaps. 3, 4, 5. We have chosen to divide the rogue wave occurrence mechanisms into (i) quasilinear ones (that usually are efficient in different geographical conditions with minor modifications, Chap. 3), (ii) nonlinear ones in water of infinite and finite depths (Chap. 4) and (iii) nonlinear ones in shallow water (then the specific wave dispersion and influence of the bottom may play an important role, Chap. 5). The essential physics of the processes of wave focusing by different mechanisms is generally well understood but their occurrence in the ocean is poorly documented. That is why we start Chaps. 3, 4, 5 with theory, modeling, and a description of the physical mechanisms followed with available testimonies of manifestations of this physics in laboratory tanks and nature.

In the Conclusion, we emphasize that most of the developed theories are applicable to other physical phenomena starting from ocean waves of different nature (wind waves, tsunamis, edge and Rossby waves) and ending with nonlinear optics (for instance optical rogue waves in fibers) and astrophysical plasma processes. This is a great implicit benefit of the freak-wave problem exploration, since rogue waves motivated significant development of nonlinear wave theories, including integrable systems and the study of instabilities, higher-order statistics, and rediscovering physical effects in new applications, etc.

This book is designed for Master and PhD students, as well as researchers and engineers in the fields of nonlinear waves, fluid mechanics, physical oceanography, ocean and coastal engineering, and applied mathematics. In Chap. 2, the fundamental basis and tools that are needed to understand and analyze the various mechanisms generating the extreme wave events given in Chaps. 3, 4, 5 are presented. For a deeper knowledge of some specific methods, the reader can refer to the bibliography, which is well stocked with references.

Marseille, France *Christian Kharif*
Nizhny Novgorod, Russia *Efim Pelinovsky*
Nizhny Novgorod, Russia *Alexey Slunyaev*

Acknowledgments

The authors would like to acknowledge the Centre National de la Recherche Scientifique (CNRS) and the Ecole Centrale de Marseille (ECM) for their support. In the final stages, the authors' work was supported by grant INTAS 06-1000013-9236. Efim Pelinovsky and Alexey Slunyaev also acknowledge support from the Russian Foundation for Basic Research (RFBR) (06-05-72011, 08-02-00039, 08-05-00069). The research activity of Alexey Slunyaev was also supported by the grant of the President of Russian Federation MK-798.2007.5.

We would like to thank all our colleagues and coauthors who helped build a better comprehension of rogue wave physics. Namely, D. Clamond, I.I. Didenkulova, M. Francius, J.P. Giovanangeli, J. Grue, A.A. Kurkin, B.V. Levin, L.I. Lopatukhin, A.V. Sergeeva, C.G. Soares, T. Soomere, T.G. Talipova, and J. Touboul.

The authors are grateful for the stimulating discussions, and provided materials and photos to S. Haver, I.V. Lavrenov, O. Kimmoun, A.B. Rabinovich, M. Sokolovsky; and K. Hutter for helping us to improve the text of the manuscript.

Finally, we would like to express our appreciation for the understanding and support from our families.

Contents

Introduction 1
    References 9

1 Observation of Rogue Waves 11
    1.1 Historical Notes and Modern Testimonies 11
    1.2 Instrumental Registrations and Related Problems 20
        1.2.1 Keystones of the Rogue Wave Measurements 20
        1.2.2 Time-Series with Rogue Wave Occurrence 22
        1.2.3 SAR Registrations of Rogue Waves 26
    1.3 Sea States 26
    References 29

2 Deterministic and Statistical Approaches for Studying Rogue Waves . 33
    2.1 Deterministic Equations 34
        2.1.1 Mass and Momentum Conservation Equations 34
        2.1.2 Boundary Conditions 37
        2.1.3 Linearization: Equations for Small Amplitude Waves .. 38
        2.1.4 Dispersion Relation 40
    2.2 Statistical Description 43
        2.2.1 The Rayleigh Probability 44
        2.2.2 Wave Spectra 48
        2.2.3 Kinetic Models 53
    2.3 Possible Physical Mechanisms of Rogue Wave Generation 56
        2.3.1 Wave-Current Interaction 57
        2.3.2 Geometrical or Spatial Focusing 58
        2.3.3 Focusing Due to Dispersion: The Spatio-Temporal Focusing 58
        2.3.4 Focusing Due to Modulational Instability 58
        2.3.5 Soliton Collision 58
    References 60

3 Quasi-Linear Wave Focusing 63
    3.1 Geometrical Focusing of Water Waves 63
    3.2 Dispersive Enhancement of Wave Trains 69

　　　　3.2.1　Exact Solution for the Delta-Function 72
　　　　3.2.2　Exact Solution for a Gaussian Wave Train 73
　　3.3　Wave Focusing Under the Action of Wind 78
　　3.4　Wave-Current Interaction as a Mechanism of Rogue Waves 81
　　References 87

4　**Rogue Waves in Waters of Infinite and Finite Depths** 91
　　4.1　The Modulational Instability 92
　　　　4.1.1　Within the Framework of the Fully Nonlinear Equations ... 92
　　　　4.1.2　Within the Framework of the Nonlinear
　　　　　　　Schrödinger (NLS) Equation 95
　　4.2　Rogue Wave Phenomenon within the Framework of the NLS
　　　　Equation 105
　　　　4.2.1　General Solution of the Cauchy Problem 106
　　　　4.2.2　Nonlinear-Dispersive Formation of a Rogue Wave 107
　　　　4.2.3　Solitons on a Background and Unstable Modes 111
　　4.3　Rogue Wave Simulations within the Framework of the Fully
　　　　Nonlinear Equations 117
　　　　4.3.1　A High-Order Spectral Method 117
　　　　4.3.2　A Boundary Integral Equation Method 120
　　　　4.3.3　Numerical Simulation of Rogue Waves Due
　　　　　　　to Modulational Instability 121
　　　　4.3.4　Numerical Simulation of Rogue Waves Due to Dispersive
　　　　　　　Focusing in the Presence of Wind and Current 130
　　　　4.3.5　Numerical Simulation of Rogue Waves Due
　　　　　　　to Envelope-Soliton Collision 135
　　4.4　Statistical Approach for Rogue Waves 140
　　4.5　Laboratory Experiments of Dispersive Wave Trains
　　　　with and without Wind 143
　　4.6　Three-Dimensional Rogue Waves 147
　　4.7　In Situ Rogue Waves 154
　　　　4.7.1　Nonlinear Analysis of Measured Rogue Wave Time Series .. 155
　　　　4.7.2　Statistics from Registrations of Natural Rogue Waves 162
　　References 165

5　**Shallow-Water Rogue Waves** 173
　　5.1　Nonlinear Models of Shallow-Water Waves 173
　　5.2　Nonlinear-Dispersive Focusing of Unidirectional Shallow-Water
　　　　Wave Fields 177
　　5.3　Numerical Modeling of Irregular Wave Fields in Shallow Water
　　　　(KdV Framework) 184
　　5.4　Three-Dimensional Rogue Waves in Shallow Water 191
　　5.5　Anomalous High Waves on a Beach 198
　　　　5.5.1　Waves at Vertical Walls 199
　　　　5.5.2　Wave Run-up on a Plane Beach 203
　　References 207

6	**Conclusion** 211	
	References 212	
A	**Discretisation of the Boundary Integral Equation for the Potential** ... 213	

Index 215

Introduction

In this section, the matter of the problem and general views are discussed. We highlight the facts that made people realize that there was a problem, and discuss the main questions surrounding the phenomenon of rogue waves.

> "Our captain, who has 20 years on the job, said he never saw anything like it."
> — Susan Robison, Norwegian Cruise Line spokeswoman, New York Daily News, April 17, 2005

There are a number of well-documented cases of the occurrence of unexpectedly large waves; some of them are described in Chap. 1, and other descriptions may be found in references therein. It is well understood that the sea may be dangerous for sailing. It is also generally recognized that the modern level of engineering is high and can generally protect people from many disasters. But where does the problem lie? People are accustomed to thinking that the construction and technical equipment of modern ships can allow safe sailing everywhere on the ocean. This confidence might be true if we had a full and realistic comprehension of all the possible dynamics on the sea surface, but this is not true.

The first vital question arises about the possible maximum wave heights on the sea surface generated by the wind. The wave height H is defined as the vertical distance between the wave crest and the deepest trough preceding or following the crest (see Fig. I.1, and (Massel 1996) for details).

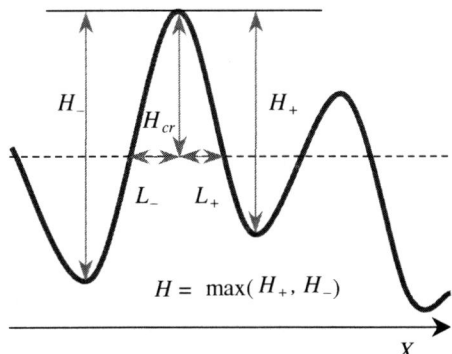

Fig. I.1 A cross section of a sea surface wave profile propagating in X direction

C. Kharif et al., *Rogue Waves in the Ocean*, Advances in Geophysical and Environmental Mechanics and Mathematics, DOI 10.1007/978-3-540-88419-4_1,
© Springer-Verlag Berlin Heidelberg 2009

When Captain Dumont d'Urville, a French scientist and naval officer in command of an expedition in 1826, reported encountering waves up to 30 meters height, he was openly ridiculed. Three of his colleagues supported his estimate but could not help him to be believed. Apparently the largest reported wave in the open sea reached a height of about 34 m (112 ft). The United States Ship (USS) Ramapo in the North Pacific reported it in 1933 (Draper 1964, Dennis and Wolff 1996). Crew members standing on the ship's bridge could measure the height of a wave by lining up its crest with the horizon and a point on the ship's mast (making the line of sight approximately horizontal) while the stern of the ship was at the bottom of a trough (see Fig. I.2).

Until now, the largest reliable instrumentally measured waves have had heights of 30 m; they were registered during the "Halloween Storm" in 1991 and Hurricane Luis in 1995. Waves with heights a little bit more than 29 m were measured under severe, but not exceptional, wind conditions in 2000 by a British oceanographic research vessel near Rockall, west of Scotland (Holliday et al. 2006). Liu and MacHutchon (2006) report higher waves, but they agree that some of them *must* be errors in the gauge, thus making the results suspect.

Nowadays, observations and measurements of high waves from space have become possible. A three-week registration of surface waves from the European satellite ERS-2 revealed regions with high waves (see Fig. I.3) and detected a wave of 29.8 m height. Bearing in mind that ships are often designed for 10–15 m wave heights, it becomes obvious that the observed waves are real threats that may cause damage and even the loss of ships (Faulkner 2001).

High waves are usually generated by storms and hurricanes; and rogue waves are obviously also much more probable during severe weather (Guedes Soares et al. 2004). Komar (2007) reports of a substantial increase in typical wave heights during a season of tropical storms and hurricanes in the North Atlantic. The rate of increase for one of the buoys used in the study is 5.4 cm per year, which has resulted in 1.8 m growth for the period of 1975–2005. The most likely explanation for that it is related to the progressive intensification of the hurricanes themselves.

Most of the casualties (about 60%) are related to operational causes (e.g., fire, collision, machinery damage), while the remaining 40% are characterized by design and maintenance causes (i.e., water ingress, hulls breaking into two pieces, and capsizing). In the case of marine structures (such as oil and gas platforms), the role of the design is even more important since a platform cannot tack, and meets

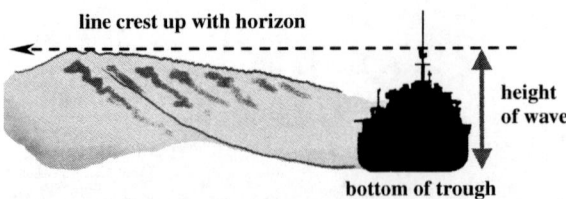

Fig. I.2 Observation of the highest reported wave by the crew members of the United States Ship "Ramapo" (Dennis and Wolff 1996)

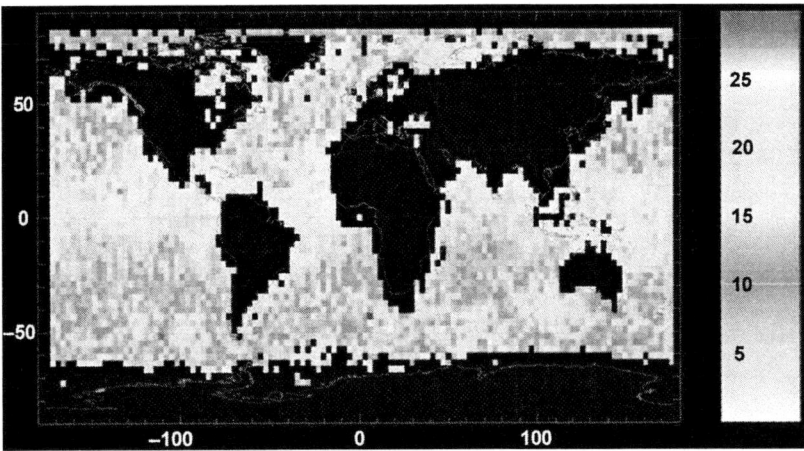

Fig. I.3 Map showing maximum single wave heights (in meters) derived from three weeks of ERS-2 SAR data acquired in August-September 1996. Reproduced from (Rosenthal et al. 2003)

a wave "as it is." Practical designs always involve compromises between safety and efficiency, and the goal is to account for expected events over the useful lifetime of a ship or structure. The crucial question that should be answered when estimating the danger is how often extreme events actually happen.

For example, the present Norwegian Petroleum Directorate's regulations describe that loads in the ultimate limit state and the serviceability limit state controls should be checked with an annual probability of 10^{-2} (once in 100 years). These waves may hit the deck structure, but they should not cause damage; the platform should be capable of full operation after an incident. The waves should not hit areas where people can be hurt. Imposing restrictions for personnel in certain areas can meet this last requirement. Loads in the accidental limit state control should meet an annual probability rate of 10^{-4} (once in 10,000 years). The total safety of the platform should not be jeopardized, personnel should have the possibility to be safely evacuated, and no major pollution should occur. Localized damage during a severe storm does not necessarily mean that a platform was poorly designed. Occasional damage might be repaired at a lower cost than building and installing a platform with a higher deck.

The current state of affairs, however, is obviously not acceptable. Casualties happen too frequently and are too dramatic. Hundreds of vessels sink and hundreds of people perish annually (see Fig. I.4), although the situation has taken a turn for the better over the last few years. The list of accidents related to the attacks of huge waves contains many recent dates. Twenty-two (22) super carriers were lost or severely damaged between 1969 and 1994 due to the occurrence of sudden rogue waves; a total of 542 lives were lost as a result (Lawton 2001). About 650 incidents are counted during the period from 1995 to 1999 due to bad weather, including total losses of all propelled sea-going merchant ships in the world weighing 100 gross tons or more (see Fig. I.5). Thirty-six percent (36%) of them foundered, 25% suffered water ingress, 6% incurred evere hull damage, and 8% capsized as intact ships (Toffoli et al. 2005).

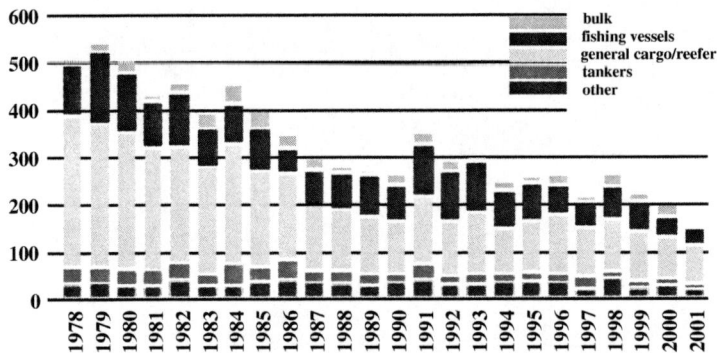

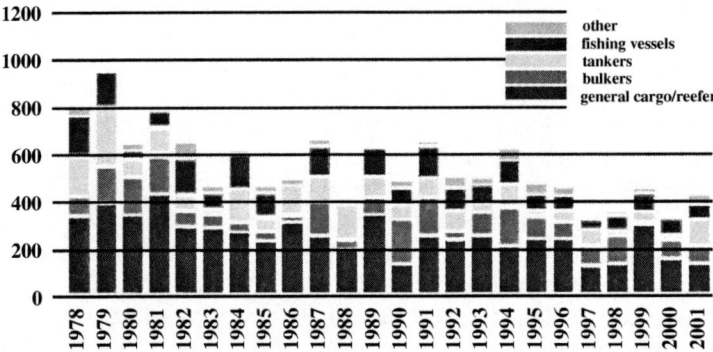

Fig. I.4 Number of total losses and number of fatalities per year of crew and passenger during 1978–2001 (Source: Det Norske Veritas, http://www.dnv.com/)

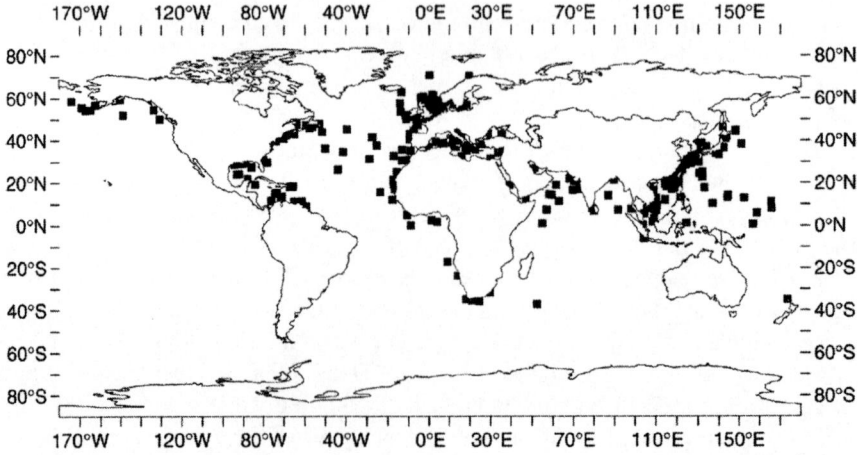

Fig. I.5 Distribution of shipping accidents from 1995–1999. (Toffoli et al. 2005, reproduced with permission from Elsevier)

Offshore platforms are also vulnerable to rogue waves. On 15 February 1982, a giant wave smashed the windows and flooded the control room in a drilling rig run by Mobil Oil on the Grand Banks of Newfoundland. Shortly afterwards the rig capsized and sank, killing all 84 people on board (Lawton 2001). The famous New Year Wave attacked the Draupner Jacket platform on 1 January 1995, with a height close to 26 m while the typical surrounding waves were about 11–12 m and the maximum expected wave height was estimated at about 20 meters (Karunakaran et al. 1997, Trulsen and Dysthe 1997).

The number of accidents reported by the mass media is growing, and the problem of huge sea waves has attracted many people's attention. Striking photos of damage collected in Fig. I.6 prove that those waves were really abnormal for the ship design of the time. Recent accidents with large passenger carriers (Queen Elizabeth 2 in 1995, Caledonia Star and Bremen in 2001, and Explorer, Voyager, and Norwegian Dawn in 2005) demonstrate the potential threat of rogue waves to normal people, while casualties with a subsequent pollution of large coastal areas (Erika in 1999, Prestige in 2002) show examples of indirect losses and the importance of safe navigation on a global scale.

So, the importance of the safe use of ocean stationary and drifting structures is obvious, as well as the message that current theoretical and engineering models underestimate the occurrence of extreme sea waves.

Fig. I.6 Photos of damage caused by huge waves (from Olagnon (2000))

Two different types of waves usually characterize the sea surface on a scale of a few meters to a few hundred meters. They are associated with wind above waves: wind waves and swells. Whereas the first refers to waves still under the influence of the wind, the latter refers to waves that have already moved out of the generating area or are no longer affected by the wind. The relatively frequent occurrence of freak wave events and the spreading of these accidents throughout the world's oceans (see Fig. I.7) allows us to believe that the freak wave phenomenon is related to the dynamics of typical waves on the sea surface—i.e., generated by the wind and more or less freely propagating.

The "wave age"[1] may be characterized by the distance (fetch) over which the wind blows over the sea surface. Various wave amplification mechanisms have been suggested by different authors (see Belcher and Hunt 1993). Due to the gravity force, the surface perturbations split into traveling waves. Qualitatively, the fully developed waves (with a long fetch, which needs large areas) depend on the wind speed only. According to dimensional analysis, the wave periods are then expressed as $T \sim U_w/g$, where U_w is the wind speed and $g = 9.8$ m/s^2 is the gravity acceleration. Thus, the stronger the wind is, the longer the waves will be. The surface waves have periods of several seconds in weak wind, 8–10 s in moderate wind, and 20–30 s in very strong winds. Free gravity surface waves over the deep ocean have a phase speed of $C_{ph} = gT/(2\pi)$ (see details in Chap. 2), and therefore the wave lengths $\lambda = C_{ph}T$ vary from several meters up to several hundred meters. In comparison with wind seas, swells generally have longer periods and larger lengths.

Small-amplitude waves are almost sinusoidal, although large-amplitude waves are not symmetric due to nonlinear bound wave corrections. Because of this effect

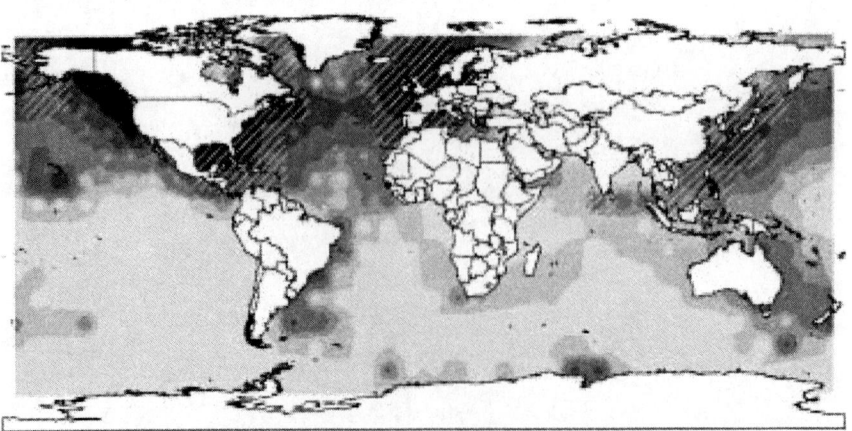

Fig. I.7 Global distribution of ship density (intensity of the gray color) and locations of accident occurrences (*hatched*). (Monbaliu and Toffoli 2003, reproduced with permission)

---

[1] More exactly, the wave age is defined as the ratio C_{ph}/U_{10} or C_{ph}/U_* where C_{ph} is the phase speed of water wave components at the spectral peak frequency and U_{10} and U_* are the wind velocity at height 10 m above the mean level and friction velocity respectively.

the crests become sharper, while troughs – smoother. Waves cannot be too high. Due to nonlinearity they break. In the open sea (when water depth much exceeds the wavelengths) the strength of nonlinearity is characterized by the wave steepness $s = KH/2$, where H is the wave height already introduced, and $K = 2\pi/\lambda$ is the wavenumber. In most cases, a regular plane wave (i.e., a wave that has a permanent profile in the crosswise direction) comes to the breaking onset when the steepness has a value of about $s \approx 0.4$. Thus, a 30 m breaking wave has a length of about 250 m and a period of about 12 s. These wave estimations look quite realistic.

The breaking phenomenon restricts the wave heights. Young waves are shorter than old ones. For short-fetch situations, growing waves are inhibited by breaking before they can grow very high. This view is supported by observations that typical waves do indeed tend to break in developing seas while smaller-scale waves tend to break in fully-developed seas. Rather large mean wave steepness is often reported in areas of relatively low significant wave height.

On the whole, the global wave climate indicates that high-wave activities are located at the highest/lowest latitudes (Fig. I.3). Ocean regions such as the North Pacific and the North Atlantic, the North Sea, the Gulf of Alaska, and the Bering Sea show the most severe sea states. However, the largest significant wave height does not occur necessarily where the largest wave steepness occurs. High steepness was reported close to the eastern coast of North America, the southern North Sea, the Mediterranean Sea, and the eastern coast of Asia, where the significant wave height was often lower than 3 m (Monbaliu and Toffoli 2003, Toffoli et al. 2005).

Relatively high waves are expected to be recorded during specific incidents. Toffoli et al. (2005) found, however, that rather low significant wave heights occurred during certain ship accidents that were reported as being due to bad weather. Thus, we are forced to come to the conclusion that wave height is not the only significant injurious factor that gives waves rogue status.

Indeed, the wave impact upon marine structures may be determined by other parameters, such as steepness, crest height (H_{cr}), and horizontal wave asymmetry (difference in L_+ and L_-) (see Fig. I.1), etc. Different types of ships may suffer from different wave parameters and conditions. Toffoli et al. (2005) note, for example, that fishing vessels have mainly capsized while fishing or loading fish. This is an important practical question that is not fully answered.

On the other hand, existing measurements and theories do not always allow a very detailed description of the accidents. Thus, a simplified definition of a freak wave becomes relevant. In this book, we employ the simple definition that a freak wave exceeds at least twice the significant wave height:

$$AI > 2, \quad \text{where} \quad AI = \frac{H_{fr}}{H_s}. \tag{I.1}$$

Here, H_{fr} is the height of the freak wave, and H_s is the significant wave height, which is the average wave height among one third of the highest waves in a time series (usually of length 10–30 min). In that way, the abnormality index (AI) is the only parameter defining whether the wave is rogue or not.

An alternative point of view exists that there are *rogue waves* that consist of two populations: (i) *"classical" extreme waves* (that are described by conventional physics, models and statistics) and (ii) *"freak" extreme waves* (that need new approaches and theories) (Haver 2005). This concept is based on probabilistic considerations. In this book, we are more interested in physical mechanisms and statistics of all kinds of extreme waves, thus we do not make such separation and consider all terms listed in the Preface (rogue, freak, etc. waves) to be synonyms and applicable to a wave if it agrees with condition (I.1). Doing a simple statistical analysis of the Reference Lists of this book, one can easily see that the word "rogue" may be found there most frequently, "freak" is less frequent, and "extreme" is at the bottom of this popularity rating. This may support (in part) the title of the book, where the term "rogue" is used instead of all others.

Hundreds of waves satisfying condition (I.1) have been recorded by now (see Chap. 1), and several waves with an abnormality index larger than three ($AI > 3$) are known. Theoretical predictions allow even higher rates of wave amplification. This is seemingly confirmed by the results of Liu and MacHutchon (2006); they hypothesize that "typical" rogue waves achieve amplification in the range of $2 < AI < 4$. Nevertheless, the variety of conditions when the waves were measured do not allow for rigorous statistical study of these waves—they still remain exceptional events.

There are a number of questions that arise and need to be answered—some of them are given here and many are the titles of recent scientific articles:

- Are there different kinds of rogue waves?
- Are rogue waves beyond conventional predictions?
- Are new physics really necessary?
- Freak waves – rare realizations of a typical extreme wave population or typical realizations of a rare extreme wave population?
- Are extreme waves the largest ever recorded?
- Were freak waves involved in the sinking of [this or that ship]?
- Are rogue waves a problem for structural design?
- Are there particular oceanographic conditions in which freak waves are more probable?
- Do extreme waves appear in groups (the "Three (nine) Sisters" of mariners' lore)?
- Can a "wall of water" be spotted enough in advance to allow time for safety measures?
- Can one identify and track a group within which a rogue wave might suddenly appear?
- Modeling a "rogue wave" – speculations or realistic possibility?
- What factors limit extreme wave heights?
- Can the Benjamin-Feir instability spawn a rogue wave?
- Rogue waves and wave breaking – how are these phenomena related?
- What effect does the wind produce on the kinematics and dynamics of rogue waves?

The purpose of this book is to show the progress that is being made in approaching the answers in the list above as well as other questions, and to consider some new questions that should be answered in the future. The main attention will be focused on the physical mechanisms of rogue wave generation brought into correlation with experiments and natural observations.

List of Notations

AI	abnormality index
C_{ph}	phase velocity
g	acceleration due to gravity
H	wave height
H_{cr}	wave crest height
H_{fr}	height of the freak wave
H_s	significant wave height
K	wavenumber
S	wave steepness
T	wave period
U_w	wind velocity
X	coordinate along the wave propagation
λ	Wavelength

References

Belcher SE, Hunt JCR (1993) Turbulent shear flow over slowly moving waves. J Fluid Mech 251:109–148
Dennis J, Wolff G (1996) Waves. Freak waves and Rogues. In: The bird in the waterfall. http://www.leelanau.com/waterfall/soundandfury.html. Accessed 13 March 2008
Draper L (1964) 'Freak' ocean waves. Oceanus 10:13–15
Faulkner D (2001) Rogue waves – defining their characteristics for marine design. In: Olagnon M, Athanassoulis GA (eds) Rogue Waves 2000, Ifremer, France, pp 3–18
Guedes Soares C, Cherneva Z, Antão EM (2004) Abnormal waves during Hurricane Camille. J Geophys Res 109:C08008. doi:10.1029/2003JC002244
Haver S (2005) Freak waves: a suggested definition and possible consequences for marine structures. In: Olagnon M, Prevosto M (eds) Rogue Waves 2004, Ifremer, France
Holliday NP, Yelland MJ, Pascal R et al (2006) Were extreme waves in the Rockall Trough the largest ever recorded? Geophys Res Lett 33:L05613. doi:10.1029/2005GL025238
Karunakaran D, Bærheim M, Leira BJ (1997) Measured and simulated dynamic response of a jacket platform. In: Proc 16th Symp OMAE 1997, Yokohama, Japan, 1997, vol II:157–164
Komar PD (2007) Higher waves along U.S. East coast linked to hurricanes. Eos Trans AGU 88:301
Lawton G (2001) Monsters of the deep (The Perfect Wave). New Scientist 170 No 2297:28–32
Liu PC, MacHutchon KR (2006) Are there different kinds of rogue waves? In: Proc 25th Int Conf OMAE 2006, Hamburg, Germany, 2006, OMAE2006-92619:1-6
Massel SR (1996) Ocean surface waves: their physics and prediction. World Scientific Publishing Co Pte Ltd, Singapore

Monbaliu J, Toffoli A (2003) Regional distribution of extreme waves. In: Rogue Waves: Forecast and Impact on Marine Structures. GKSS Research Center, Geesthacht, Germany

Olagnon M (2000) Vagues extrêmes – Vagues scélérates. http://www.ifremer.fr/web-com/molagnon/jpo2000/. Accessed 13 March 2008

Rosenthal W, Lehner S, Dankert H et al (2003) Detection of extreme single waves and wave statistics. In: Rogue Waves: Forecast and Impact on Marine Structures. GKSS Research Center, Geesthacht, Germany

Toffoli A, Lefevre JM, Bitner-Gregersen E, Monbaliu J (2005) Towards the identification of warning criteria: Analysis of a ship accident database. Appl Ocean Res 27:281–291

Trulsen K, Dysthe KB (1997) Freak waves—a three-dimensional wave simulation. In: Proc 21st Symp on Naval Hydrodynamics. National Academy Press, USA, 550–560

Chapter 1
Observation of Rogue Waves

There are a number of personal descriptions of unexpectedly high waves collected in the literature by now. Some of them will be discussed hereafter. Besides the reports, there also exist some dilettante photos of rogue waves; many of them may be found on the Internet. Instrumental measurement is a more substantial kind of finding evidence of freak waves. They are made by gauges of different types and may be used for validating theories and models and for reproducing the events in laboratory experiments. The overwhelming majority of the available instrumental records represent time series of the values of surface elevation (made by buoys or altimeter gauges). Three-dimensional (3D) records (and especially their sequences) of surface waves made by space or airborne synthetic-aperture radar (SAR) are recent data containing the most complete information about the waves. The latter measurements are not very well validated at present (retrieving sea surface elevation fields from "imagettes"). At the same time, personal observations may be useful since they contain qualitative information about the 3D wave structure and its dynamics. Some of these descriptions—historical and recent testimonies—are collected in Sect. 1.1. Section 1.2 is dedicated to the instrumental records of rogue waves: a survey of available rogue wave records, techniques of wave measurements, and problems of reliability of the high-wave measuring technique. Section 1.3 classifies the sea states and shows their relation to rogue wave occurrence.

1.1 Historical Notes and Modern Testimonies

Personalities make history human. Our story is created by accidents. The freak wave phenomenon could remain marine folklore if there were no crashes that shake people's minds. Notorious casualties attract attention to the existence of abnormally huge waves, and evidence makes us believe the reports. A long but obviously incomplete list of accidents starting from the time of Christopher Columbus has been collected by Liu (2007). Many other descriptions are available in various publications (Mallory 1974, Torum and Gudmestad 1990, Haver and Andersen 2000, Lawton 2001, Olagnon and Athanassoulis 2001, Kharif and Pelinovsky 2003) and references therein. The stories are sometimes very similar, but frequently they show distinctive differences and may be useful for the comprehension of the phenomenon. We represent below some stories describing different kinds of rogue wave accidents.

The most striking cases of rogue waves correspond to **strongly localized high waves**.

> "Down the ways at Quincy, Mass, last week went the largest cargo vessel ever built in the U.S., and the largest tanker in the world: the 45,130-ton World Glory, with a capacity of 16.5 million gals – enough to fill 2,062 railroad tank cars..."

This is the beginning of the history of the tanker "World Glory," announced by a newspaper in 1954 (Time 1954). Its end is not so enthusiastic. On June 13, 1968, travelling along the South African coast under the Liberian flag, World Glory encountered a freak wave, which broke the tanker into two pieces and led to the death of 22 crew members (Lavrenov 2003) (Fig. 1.1a). It happened in the Indian Ocean, 105 km east of Durban. As a result, about 14 million gallons of oil spilt into the Ocean.

The tanker Prestige (42,000 gross tons, and about 250 m in length) went down similarly off the Spanish coast in 2002 (Fig. 1.1b). Estimations of the amount of spilt oil are different, but they are roughly about 20 million gallons. Some people connected with the accident think that the damage that led to its sinking might have been caused by a freak wave. Anyway, it is more or less obvious that the hull was unable to bear the wave force. The Prestige was built more than 20 years after World Glory. The vessel met all American Bureau of Shipping Rule structural requirements and International Association of Classification Societies Rule hull girder strength requirements. The vessel was properly loaded and had adequate hull strength for the reported conditions at the time of the casualty (ABS 2003).

The number of accidents that occurred with wavelengths less than half the ship's length is small (Toffoli et al. 2005), so we could suppose that the damage in both cases was probably caused by intense long waves causing unexpected nonuniform loads on the hulls.

The cruise liner Queen Elizabeth II encountered a rogue wave in the North Atlantic about 30 m height during a storm in 1995. The ship master referred to a particular episode where they had been looking at **a wall of water** from the bridge for a couple of minutes before it hit the ship well above the waterline: *"a great wall of water – it looked as if we were going into the White Cliffs of Dover."* A similar description was given by one of the crew members of the Statoil floating rig Veslefrikk B (it was hit the same year by a wave that resulted in significant damage) (Haver and Andersen 2000). The first mate of the oil tanker Esso Languedoc described the wall of water in the photo in Fig. 1.1c (see also Fig. 1.2b): *"We were in a storm and the tanker was running before the sea. This amazing wave came from the aft and broke over the deck. I didn't see it until it was alongside the vessel but it was special, much bigger than the others."* (Lawton 2001).

Freak events represented by several successive very high waves in **wave groups** are also well known. A collision of the naval ship Jeanne d'Arc with the Glorious Three in 1963 was described in (Moreau et al. 2005).

(a)

(b)

(c)

Fig. 1.1 Accidents with huge waves. (**a**) Sinking of World Glory tanker in 1968, the photo is taken from (Liu 2007). (**b**) Sinking of tanker Prestige in 2002 (Lechuga 2006, Reproduced with permission). (**c**) This picture was taken on the oil freighter Esso Languedoc outside the coast of Durban by P. Lijour, South Africa 1980 (Reproduced from Dysthe et al. 2005). (**d**) The map of the incidents off the Southeast coast of Africa, and the scheme of the collision of tanker

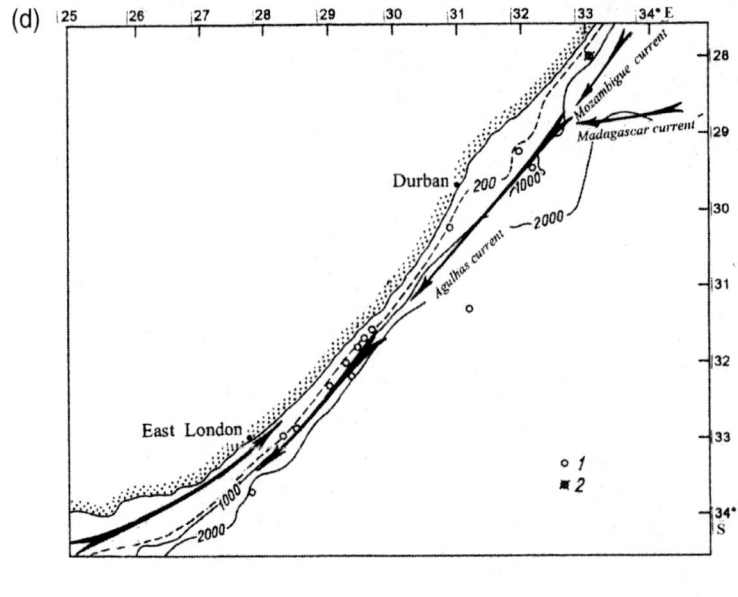

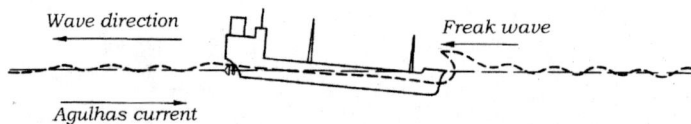

Fig. 1.1 (continuned) Taganrogsky Zaliv with a rogue wave (Reproduced from Lavrenov 2003). (**e**) A "diving" into a wave boat. The case looks similar to the descriptions of the accident with the Taganrogsky Zaliv (Reproduced from Heavy Seas 2002). (**f**) Waves observed in 2006 near Kamchatka (Photo by M. Sokolovsky, http://www.kkclub.ru). (**g**) Waves and suddeen flooding in Maracas Beach (Trinidad Island, the Antilles) in 2005; (see description in Didenkulova et al. 2006). (**h**) A 2-s photo image sequence taken on the Dianna Island (Canada) (see description in Palmer 2002)

1.1 Historical Notes and Modern Testimonies

Fig. 1.1 (continued)

(h)

Fig. 1.1 (continued)

"At about 09:47 a group of large breaking waves was sighted straight ahead, just beyond an area of relative calm water (4–5 m wave height). The first wave heaved the ship; its height was estimated about 15 m. During the interval of about 100 meters in-between the first and the second wave the "Jeanne d'Arc" had time to return approximately to its waterline, but she was soon heeled over to starboard by the second wave, until the heel angle reached about 35°. During clearance of those two waves, the freeboard deck and the quarterdeck were submerged in turn, the sea covered the catwalks of the first deck, water reaching the top of the bulkheads at the time of maximum heel. The third wave was cleared in similar conditions, but with not as large amplitude motions, its height being slightly less than that of the two first ones."

Two unexpectedly large successive waves shattered windows 28 m above the waterline of the cruise liner Queen Elizabeth in 1943; two other waves capsized the trawler Kotuku in 2006; and three large waves hit and threw the fishing boat Starrigavan onto a jetty in 2007 (Lawton 2001, Liu 2007).

Splitting ocean surface waves into groups is a natural process, which will be discussed later in Chap. 4, and the central individual waves within a group are more energetic. In the past, seamen of different nationalities mention monstrous wave groups. It is interesting to note that the number of individual waves that supposedly forms a rogue wave packet is different: *three sisters* or *the ninth billow*. Surfers sometimes wait for the largest, or *seventh*, wave. Lehner (2005) notes that successive large single-wave crests or deep troughs can cause severe damage due to their impact, or may excite the resonant frequencies of the structures.

The Soviet refrigerator tanker Taganrogsky Zaliv was subjected to an abnormal wave, a **hole in the sea**, in 1985 (see Fig. 1.1d,e) (Lavrenov 2003).

> "Wave height did not exceed 5 m and the length was 40–45 m. The speed of the ship was diminished to a minimum in order to make a safer control of the ship's movement. The ship rode well on the waves. The fore and main deck were not flooded with water. At one o'clock the front part of the ship suddenly dipped, and the crest of a very large wave appeared close to the foredeck. It was 5–6 m higher over the foredeck. The wave crest fell down on the ship. One of the seamen was killed and washed overboard. Nobody was able to foresee the appearance of such a wave. When the ship went down, riding on the wave, and its frontal part was stuck into water, nobody felt the wave's impact. The wave easily rolled over the foredeck, covering it with more than two meters of water. The length of the wave crest was not more than 20 m."

Very similar descriptions are related to accidents with the cruiser Birmingham in 1944 and some other vessels (Haver and Andersen 2000). They report sighting a long trough followed by a steep crest, or a "hole" in the sea. There is a viewpoint that a hole in the sea is more dangerous for a boat than a crest, since it is less noticeable among the sea waves than huge crests, and the shipmaster cannot change course and prepare the ship in advance.

The NOAA's 56-foot research vessel Ballena capsized in an **individual rogue wave** south of Point Arguello, California in 2000. The weather was good, with clear skies and glassy swells (1.5–2 m). At approximately 11:30, the crew observed a 4.5 m swell begining to break about 30 m from the vessel. The wave crested and broke above the vessel, caught the Ballena broadside, and quickly overturned it (Kharif and Pelinovsky 2003).

Russian kayakers were lucky to observe and make photos of strange waves 25 km from Cape Olga, Kronotsky Peninsula, about 1–1.5 km offshore (Fig. 1.1f). They reported that the weather was calm with only very long gently sloping surge waves coming from the open ocean every 15–20 s. About 10 strange waves were observed in the same area with irregular lengths. Freak waves arose, propagated, and collapsed during tens of seconds and ran for about 50 m within this time. Wave heights were about 2–4 m, and typically their length along the front was about 70 m. The first photo in Fig. 1.1f is quite challenging, although the second one (taken from another aspect) looks more ordinary.

These descriptions are in some sense similar to the first kind of observations (i.e., strongly localized high waves), but the reports emphasize individual waves that propagate for some distance and are actually not surrounded by other considerable waves. They seem to be of a solitary wave variety (see Chap. 5) and are singled out for a particular case.

Extreme **coastal wave phenomena** similar to the ocean rogue waves have been noted recently. Typically, such accidents are described as a sudden brief coastal flooding or as huge waves rushing coastal structures (raised embankment or breakwaters). Two events are given in Fig. 1.1g and h; other descriptions may be found in Kurkin and Pelinovsky 2004 and Didenkulova et al. 2006. These waves have not been related with tsunamis; although it is more difficult to ascertain whether they are not caused by storm surges (this reason may likely cause the waves in Fig. 1.1g). A very high (25 m) wave splash presented in Fig. 1.1h occurred suddenly and was absolutely unexpected by the students (who made the photos) after they had spent about 45 min observing swell waves that followed a severe storm that had happened one day before.

This is a relative classification (see also Rosenthal 2005) that can be argued but cannot be finalized until all physical effects are revealed and freak wave impact is described and estimated. Some wave types are illustrated in Fig. 1.2. Different wave shapes may require different physical effects and mathematical models of different complexity to describe them. Some observers report about lifetimes of rogue events that amount to a few minutes or less.

It was already pointed out that wave height, in addition to its shape and surrounding waves may define the strength of wave impact. Unusual wavelength or small crest length (like in Fig. 1.2a) may lead to an inadmissible load distribution that may damage the hull. The most striking examples of rogue waves in the recent literature are unusually asymmetrical with high crests compared to the depth of their troughs. Presumably enormous huge-wave impacts have been already registered (Peregrine et al. 2005). Ships usually travel perpendicular to the crests with low forward speed. A particular traveling direction of a wave group results in complicated wave motion that makes the ship list and makes it difficult to safely overpass the waves. Steep waves (like in Fig. 1.2c) may yield dangerous dynamic effects due to ship motion (slamming), even though the significant wave height is not particularly large. A breaking rogue wave could potentially cause more damage than a nonbreaking wave of the same dimension. These points should be taken into account when studying the wave impact and designing a safe construction.

Due to the relatively large number of registered collisions of ships with abnormal waves, a statistical analysis of the events was performed by Toffoli et al. (2005) on the basis of 270 documented accidents selected among a total of 650 that occurred over about four years and collected by the Lloyd's Marine Information Service. Toffoli et al. (2005) emphasized that accidents occurred often in the presence of crossing seas: wind waves and swell. They claim that any significant correlation between the main surface wave parameters and ship weight were not found, although more than 90% of the incidents occurred in water depths of more than 50 m. It is suggested that different kinds of ships should be subjected to different freak-wave warning criteria.

1.1 Historical Notes and Modern Testimonies

Fig. 1.2 Rogue waves: (**a**) pyramidal wave off south Japan; (**b**) walls of water; (**c**) a very steep breaking wave crest. Reproduced from (Faulkner 2001) by permission of Ifremer, and Olagnon 2000

Observations represented in stories and even photos are insufficient for the weighty study of the rogue wave phenomenon. The suddenness of these waves results in lack of photographic evidence and sometimes confusing testimonies. Instrumental registrations provide data for quantitative analysis and careful research of physical effects underlining the occurrence of freak waves.

1.2 Instrumental Registrations and Related Problems

The history of sea elevation measurements near shore begins quite a long time ago. The first tide gauge started its record in 1806 at Brest (France). At first, these records had to track the tides, which are very important for the normal functioning of ports. Therefore, the first series were not frequently retrieved, and were not very accurate due to equipment imperfection. Shorter time scales could be measured later: tsunami waves, long sea waves. The waves in the open sea have been measured from ships with acceptable accuracy since the fifties of the XXth century, and regular research started in the sixties (see Pugh 1987, Rabinovich 1993). Nowadays, the sea-wave elevation may be measured by deployed recorders of different types, ship-, air- and spaceborne radars. The equipment is continuously being improved; new techniques perfect the instrumental observations. People obtain continuous measurements of wind-induced sea waves with the help of moored buoys and altimeters installed on platforms; these measurements represent the most useful information regarding freak-wave events.

1.2.1 Keystones of the Rogue Wave Measurements

A rogue wave is a *rare* event, and may be recorded only if long-time regular measurements are conducted. That is why the measurements performed from stationary offshore platforms and buoys are of utmost interest. At present, the number of registered freak waves is in the hundreds. Figure 1.3 shows the areas where they were continuously measured for years. All of them satisfy condition (I.1) introduced previously, although sometimes other extra conditions (such as the threshold wave height that should be exceeded by the wave) defining the freak event are applied.

A rogue period does not stand out in typical wave periods; such a wave has a period of about 10 s; the rogue event is often also quite *momentary* (not longer than a few minutes). This fact requires rather high frequency of data acquisition (in contrast to, for instance, tidal or tsunami wave recordings). It is inconsistent with the long-term character of the measurements, in the sense that the recorded data becomes enormous. Sampling with a frequency of 5 Hz represents a reasonably good resolution of the wave shape. For more than 50 thousand hours of the measurements reported in Liu and MacHutchon 2006, this results in about 10^9 single measurements. Usually data is represented by a number of 10–30 min time series. These

1.2 Instrumental Registrations and Related Problems

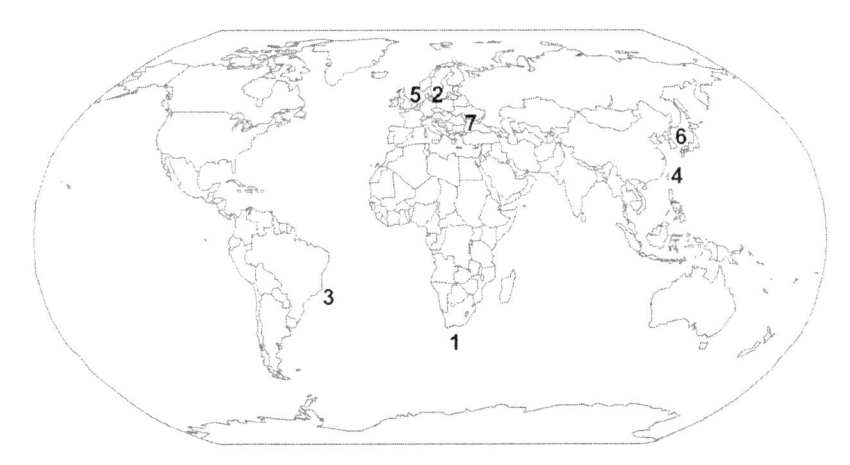

Fig. 1.3 Some instrumental registrations of freak waves (ordered by the number of reported freak waves). **1**) Offshore from Mossel Bay (1563 events, 100 m depth, gas-drilling platform) (Liu and MacHutchon 2006). **2**) The Baltic Sea (414 events, 7–20 m depth, buoys) (Paprota et al. 2003). **3**) Campos Basin near Rio de Janeiro (276 events, 1050 m and 1250 m depth, buoys) (Pinho et al. 2004). **4**) Off the eastern coast of Taiwan (175 events, 43 m depth, buoys) (Chien et al. 2002). **5**) The North Sea (at least 107 events, 126 m and 85 m depth, platforms) (Stansell 2004, 2005, Haver and Andersen 2000). **6**) Sea of Japan (14 events, 43 m depth, ultrasonic submerged gauges) (Mori et al. 2002). **7**) The Black Sea (3 events, 85 m depth, buoy) (Lopatoukhin et al. 2003, Divinsky et al. 2004)

measurements may be retrieved with some intervals (say, once an hour) if a certain condition is satisfied (a storm or high significant wave height) or may be used for processing other parameters that may be employed for the statistical study (significant and maximum wave height, wave period, etc.) and discarded afterwards. This management decreases the data volume necessary to be stored by the device.

A rogue wave is an *extreme* wave that needs an accurate and precise method of measurement. Different ways of detecting the surface elevation height are in use. They employ the reflection of an optical ray or acoustic signal by the air-sea boundary and acceleration of floatable buoys. Pressure-wave gauges may register long waves. The first type of difficulty lies in the principle of the definition of the surface elevation. The reflection of sonic or electromagnetic waves may not occur at the very air-sea boundary due to the presence of foam or bubbles that is typical in severe conditions. A buoy possesses an intrinsic moment of inertia that distorts the measurements. Jointly with the low frequency of acquisition and poor calibration, these problems may make records difficult to use in further research and unreliable.

Forristall (2005) claims that there are well-documented cases in which carefully calibrated wave recorders on the same platform give very different readings. This may result, for instance, in device errors or malfunctions, electronic noise, or interference from the structure that supports the wave sensor. Next, an error in one point of the time series may reduce the crest height to a plausible level or increase it up to an unbelievable value, if the data acquisition is not frequent. Liu and MacHutchon 2006 report huge waves, exceeding 4–10 times the significant height.

Some of the registrations may be easily rejected as spikes, but some of them cannot, since there is no provision as to how high the ratio of AI can be. In the case of buoy measurements, the actual crest height is usually underestimated (Olagnon and Magnusson 2004, Bitner-Gregersen and Magnusson 2005). Another problem is in distinguishing a very large wave from noise, which may be electronic or the result of interference from the structure that supports the wave sensor. These actual problems sometimes cast doubt on wave records and may modify results of theoretical comprehension.

1.2.2 Time-Series with Rogue Wave Occurrence

By now, thousands of measured rogue waves have been reported in the literature (see Fig. 1.3). They are the results of multiple-year measurements of surface waves. These registrations are not uninterrupted, done in different areas of the World Ocean (deep and shallow water, with and without currents), under different conditions (some registrations were performed only during storms or high significant wave heights),and by different devices. Some details are given in the figure captions or can be found in the given references. We do not discuss the measurements by Liu and MacHutchon (2006) here, since some of them are definitively just spikes.

Figure 1.4a shows a rogue wave captured in the North Sea with a record value of abnormality index, $AI = 3.19$, defined as the ratio of extreme wave to significant wave height. The famous "New Year Wave" measured on the 1st of January 1995 is shown in Fig. 1.4b. It has a very large height (about 26 m, while the amplification is more moderate: $AI = 2.24$). Haver (2005) points out that the height, however, does not exceed the so-called 100-year height, while the measured crest with height about $H_{cr} = 18.5$ m corresponds to the annual probability 10^{-4} (once in 10 000 years).

A "hole in the sea" is shown in Fig. 1.4c. Although its height is not very large, the amplification is exceptional ($AI = 2.46$). The huge wave of depression seems to be a less frequent kind of rogue wave. The intense waves are asymmetric so that the crests are typically larger than the neighbouring troughs (see Chap. 4). This can explain the prevalence of rogue crests in the amount of rogue events. Also, the time series are retrieved at one spatial point; if the lifetime of the freak event is larger than the wave period, a huge single crest should arise somewhere on the front or back of the wave.

The time series are being retrieved at one point, where the sensor is installed. When the rogue wave (or the sequence of rogue waves) is well localized in space, it will pass the registration point fast, and only one or few wave oscillations with huge amplitude will be recorded. On the other hand, the phase velocity of free waves over deep water is twice as large as the group velocity. As a result, the time series of a wave group consists of two times more individual waves than the group's snapshot.

Only a single huge wave is reported in the majority of registered rogue events. This should prove strong localization of the wave energy in space in one or a very small number of individual waves. Nevertheless "rogue groups" are also known; one example is given in Fig. 1.4d. The rogue group consists of several huge individual waves, each of which satisfies the rogue wave condition (I.1). These wave groups satisfy the nonlinear self-modulation condition (see Chap. 4) and may show different dynamics in comparison to individual freak waves, and perhaps be the consequence of another generating mechanism.

The maximum known measured wave height amplification among the data represented in Fig. 1.3 was achieved by waves measured in the Black Sea (Divinsky et al. 2004): $AI = 3.9$ (see Fig. 1.4e). The peak measurement is represented by a single point, which makes the record suspicious, but at least two other similar records exist from this buoy that show other rogue waves with somewhat less (but still large) heights that exceed significant waves. Waves presented in Fig. 1.4a–d have high resolution, and these waves are beyond any doubt.

The brief overview of known rogue wave measurements documents the existence of individual rogue crests and troughs as well as groups. Are they "pyramidal" waves or "walls of water?" This question cannot be answered on the basis of the measurements of the surface elevation in one point. The transversal effects are omitted due to the lack of single-point measurements, although it is well known that geomet-

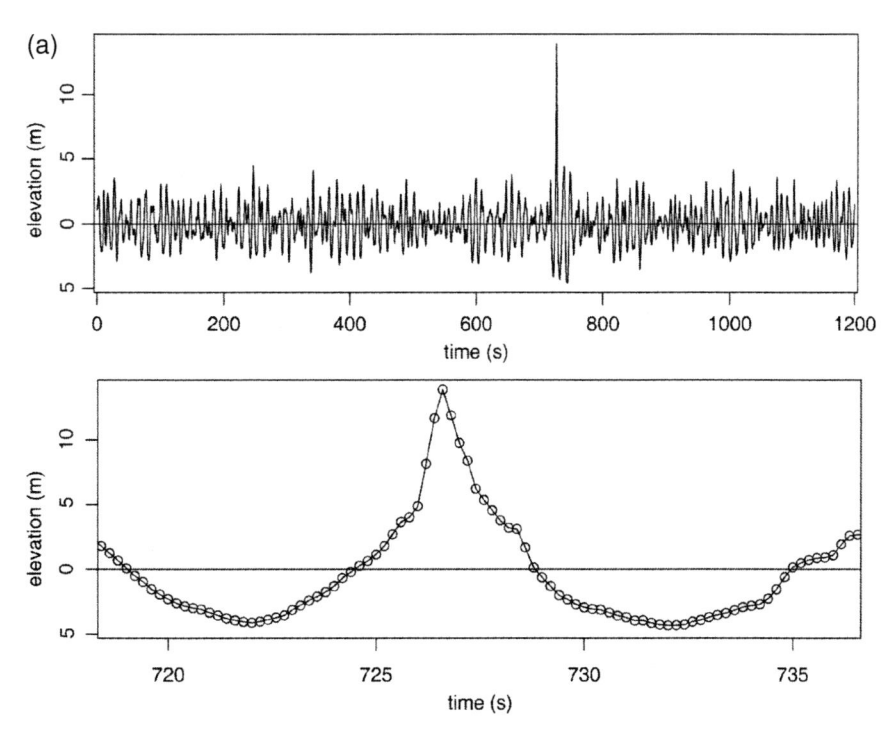

Fig. 1.4 Measured freak-wave time series. (**a**) A huge single crest (the North Sea, platform, 126 m depth, $AI = 3.19$, $H_{fr} = 18.04$ m) (Stansell 2005, reproduced with permission from Elsevier). (**b**) The "New Year Wave" (the North Sea, platform, 85 m depth, $AI = 2.24$, $H_{fr} = 26$ m). The data is granted by S. Haver. (**c**) A hole in the sea (the North Sea, platform, 126 m depth, $AI = 2.46$, $H_{fr} = 9.3$ m) (Stansell 2005, reproduced with permission from Elsevier). (**d**) A freak group (the North Sea, platform, 126 m depth, $AI = 2.23$, $H_{fr} = 13.71$ m). (**e**) A huge single crest (the Black Sea, buoy, 85 m depth, $AI = 3.91$, $H_{fr} = 10.32$ m)

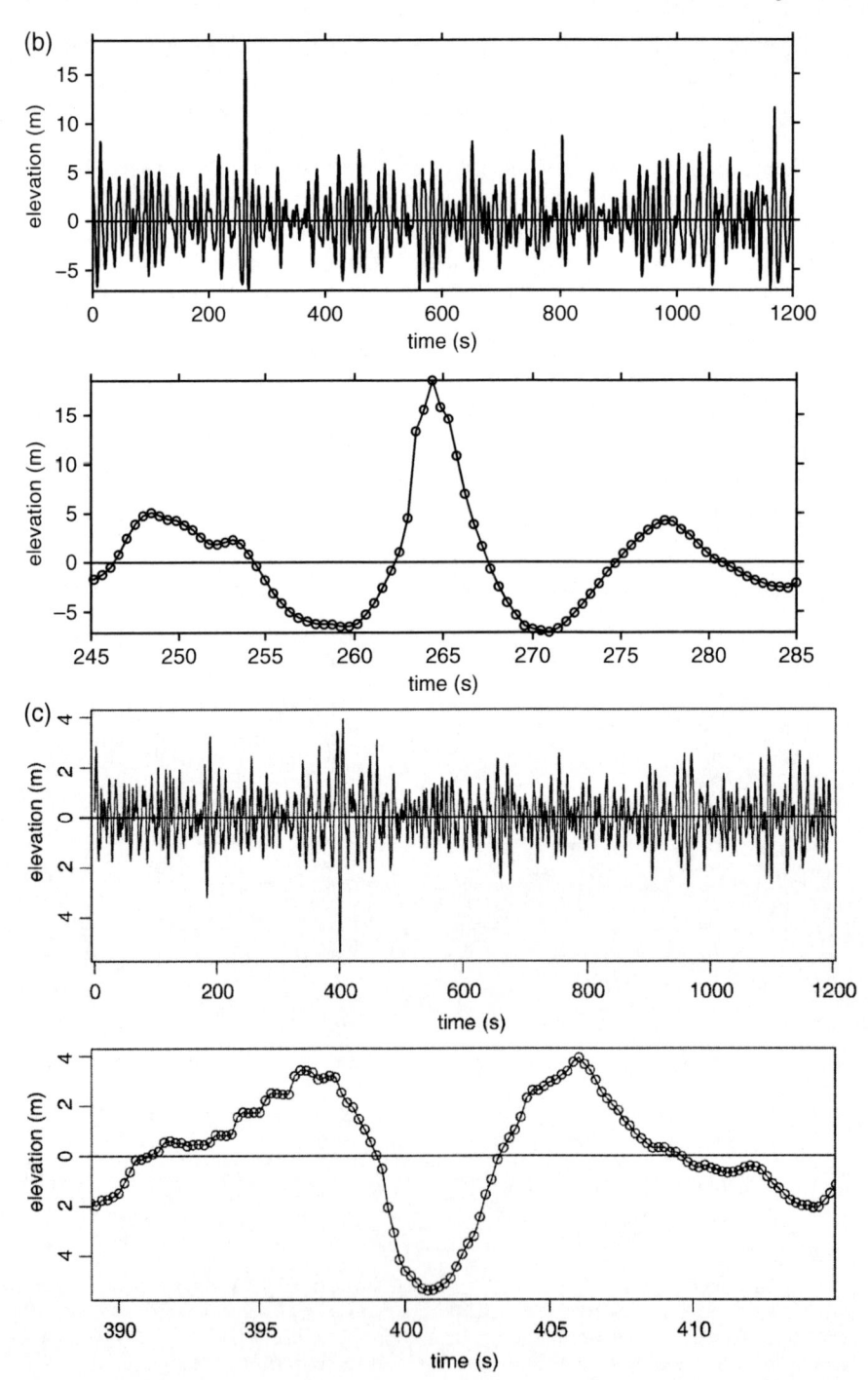

Fig. 1.4 (continued)

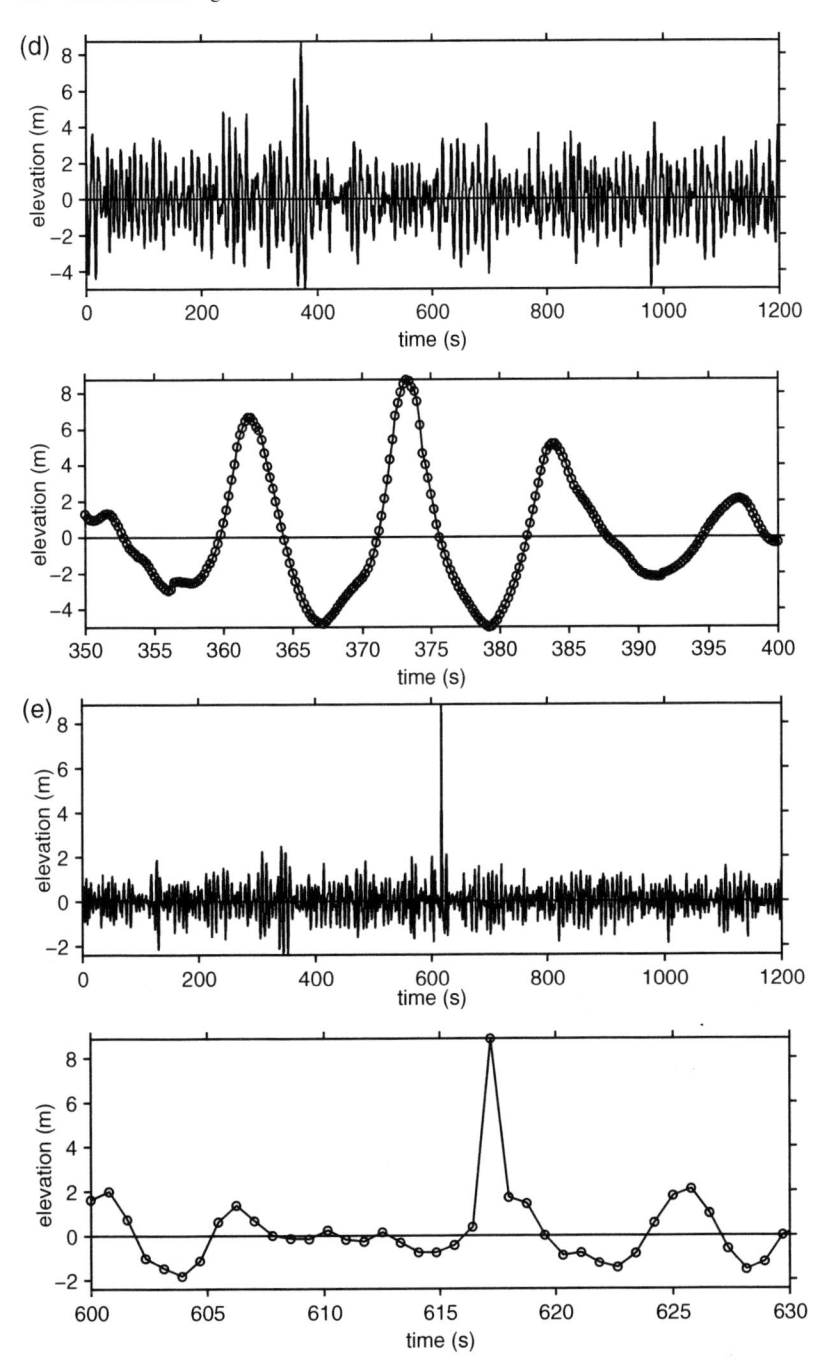

1.2 Instrumental Registrations and Related Problems

Fig. 1.4 (continued)

rical effects may play a very important role in the process of wave focusing (see Chap. 3) and significantly enhance these effects. Registrations of the 3D surface field are requested—as are their sequences—to be able to reconstruct and understand the fully dimensional surface wave dynamics. The promising approach that now is applied and developed uses synthetic-aperture radar (SAR) measurements. This technique is not recent, but needs much improvement to be able to resolve wave shapes with good accuracy (see some criticism in Dysthe et al. 2008).

1.2.3 SAR Registrations of Rogue Waves

Extreme wave events, such as rogue waves, can be detected from satellite imagery. Satellite images of the sea surface topography—including the New Year Wave on January 1, 1995—showed several extreme wave events in a 100 km × 100 km spatial domain. New spatial radar measurements (Lehner 2005, Rosenthal 2005) have been developed that allow the observation of rogue waves on a global scale. The spaceborne synthetic aperture radar, which is a high-resolution imaging system, provides images covering large areas of the sea surface, of quality sufficient to extract measurement of water waves. Spaceborne SARs on polar orbiting satellites, at approximately 800 km altitude, scan a swath of 100 km with a resolution of 20 m × 20 m at an incidence angle of 20–25° (for more details see the book by Komen et al. 1994). The physical phenomenon upon which this system is based is Bragg scattering—i.e., the resonant interaction of the incident microwaves emitted by the radar with short t. The backscattered energy is proportional to the spectral density of the short backscattering Bragg waves that depend on interactions with the long waves (Wright 1968, Valenzuela 1978, Hasselman et al. 1990). The amplitude (or energy) of the Bragg ocean waves riding on the long waves is modulated. This modulation, which is measured by SAR, allows (finally) the detection of ocean gravity waves such as rogue waves. To sum up, the long waves that modulate the short waves (Bragg waves), or their aspect, with respect to the radar will be imaged.

To obtain the two-dimensional wavenumber spectra of the surface elevation and individual sea surface topography, it is necessary to invert the radar images. Lehner (2005) investigated the behavior of single water waves, extreme waves, and wave groups by inverting SAR images into sea-surface elevations. In Fig. 1.5, there is a 5 km × 10 km normalized ERS-2 wave mode imagette acquired at 48.45° S, 10.33° E on August 27, 1996, 22:44 UTC, and the corresponding retrieved sea-surface elevation field that displays a wave of height close to 30 m.

1.3 Sea States

We have already discussed the evidence of rogue waves as an observer sees them. They are very rare events local in time (scale of seconds or few minutes) and space (scale of several wavelengths – hundreds of meters). It is quite difficult to foresee

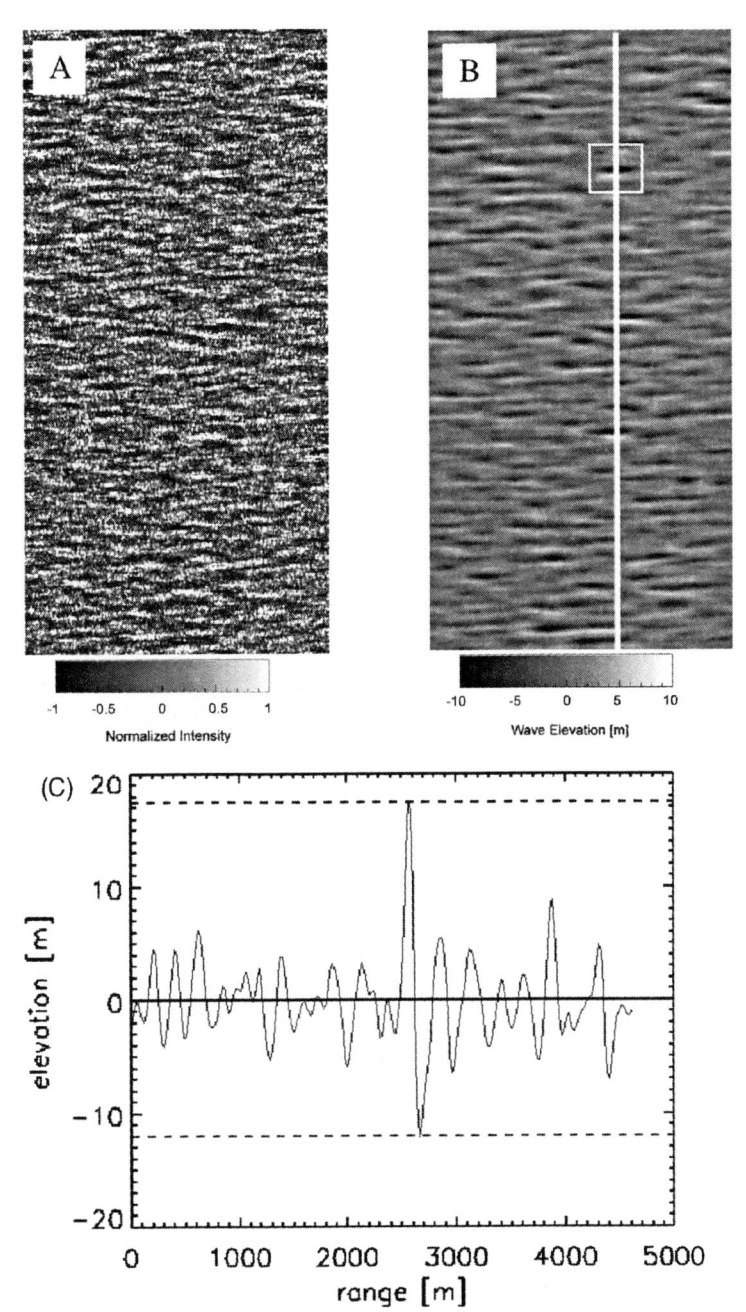

Fig. 1.5 A normalized ERS-2 wave mode imagette (**a**), the retrieved sea-surface elevation field (**b**), and the vertical transect of the retrieved ocean wave in range direction (**c**) as indicated in section (**b**) (Reproduced from Lehner 2005)

phenomena of such small scale—first, due to the difference between the scales of the forcing processes (winds, atmosphere fronts, storms, currents, and geostrophic vortices) and the surface waves; and second, due to the variety and complexity of the physical effects accompanying the wave dynamics. Meteorologists usually work with average parameters of the sea that represent the *sea state*. Significant wave height, peak period, and main wave direction are sufficient to describe sea states for the most practical purposes. It is a vital problem to manage linking the sea state characteristics with the degree of danger for navigation and sea use.

Today, this problem seems to be quite far from a solution. Lack of data and complexity of the processes prevent straightforward progress in relating the sea states and probability of rogue wave occurrence. It is more likely to reveal these dependencies on the basis of simplified models and purified conditions. Researchers fill up the lack of natural data with numerical simulation. We will consider some theoretical aspects in Chap. 2 and the results of investigations with the help of numerical modeling in Chaps. 4 and 5. Nevertheless, in this section we discuss some recent achievements in this problem due to natural data processing and analysis of the databases of the accidents.

Considering the rogue wave problem, it is first important to find the key parameters of the sea state out of the more than 100 parameters that could effectively indicate a high risk of freak waves. Many parameters are defined through the wave spectrum that will be introduced in Chap. 2. Toffoli et al. (2005) sought a correlation between ship accidents and different characteristics of sea states, such as significant wave height, mean wave period, wave steepness, and directional spread, as well as correlations between these parameters during the accidents. They report that surprisingly rather low sea states occurred during the ship accidents, while the wave parameters could reach relatively high values. This contradicts frequent conventional expectations of rogue events during significant storms. More than 50% of the incidents took place in sea states characterized by significant steepness $s > 0.1$ (where $s = K H_s/2$), although this value is not very high. They also note that relatively high values of the steepness were observed during moderate wave heights. About one half of the accident happened in crossing seas (when the wind sea and the swell directions are quite different). The higher probability of meeting a rogue wave in a crossing sea is evidently confirmed by natural data analysis in Pinho et al. 2004.

In many cases, classical parameters are unable to robustly analyze the danger of the sea state (for instance, Olagnon and Magnusson 2004, Toffoli et al. 2005) and their development in time may play an important role in forecasting. On the other hand, Bitner-Gregersen and Magnusson (2005) report that extreme events appear at different times in the storm histories—before, at, and after the significant wave height culmination. Another problem is that some sea-state parameters indicate well the presence of a rogue wave, but reach typical values just when the rogue signal disappears (or when it is just removed from the time series of the surface elevation) (Olagnon and Magnusson 2004); characteristics of this kind cannot play the role of predictors either.

The fetch that characterizes the wave development is one of the most significant parameters of the sea state. Figure 1.3 shows that on a global scale, maximum waves usually appear in vast areas. Lehner (2005) reports that the highest waves are observed when they are focused in a current, or generated in a moving fetch situation, in which the strongest wind field travels with the group velocity of the waves. Ship accidents caused by extreme waves happened mainly in crossing seas or under fast-changing weather conditions. Melville et al. (2005) claim that large waves can "pop up out of nowhere," even at small fetches (25 km); they may cause a danger to smaller vessels. Other wave parameters have been considered as candidates able to foresee the rogue wave occurrence, as it will be discussed in Chaps. 4 and 5. This search is still in progress.

Too few data sets including rogue events have been recorded, making it difficult to develop satisfactory models for the understanding and prediction of these waves. The investigations briefly collected above prove the complexity and difficulty of this problem. We are still far from being able to foresee a high probability of a rogue wave event on the basis of meteorological data. New long-term accurate measurements should be preformed to relate the sea conditions with rogue wave occurrence probability.

List of Notations

AI	abnormality index
H_{cr}	wave crest height
H_{fr}	height of the freak wave
H_s	significant wave height
K	wavenumber
s	wave steepness

References

ABS (American Bureau of Shipping) (2003) Technical analyses related to the Prestige casualty on 13 November 2002. http://www.eagle.org/news/press/prestige/ Tech_Analysis_final.pdf. Accessed 14 March 2008

Bitner-Gregersen EM, Magnusson AK (2005) Extreme events in field data and in a second order wave model. In: Olagnon M, Prevosto M (eds) Rogue Waves 2004, Ifremer, France

Chien H, Kao C-C, Chuang LZH (2002) On the characteristics of observed coastal freak waves. Coast Eng J 44:301–319

Didenkulova II, Slunyaev AV, Pelinovsky EN, Kharif Ch (2006) Freak waves in 2005. Nat Hazards Earth Syst Sci 6:1007–1015

Divinsky BV, Levin BV, Lopatukhin LI, Pelinovsky EN, Slyunyaev AV (2004) A freak wave in the Black Sea: observations and simulation. Doklady Earth Sci 395A:438–443

Dysthe K, Krogstad HE, Müller P (2008) Oceanic rogue waves. Annu Rev Fluid Mech 40:287–310

Dysthe KB, Krogstad HE, Socquet-Juglard H, Trulsen K (2005) Freak waves, rogue waves, extreme waves and ocean wave climate. http://www.math.uio.no/~karstent/waves/ index_en.html. Accessed 14 March 2008

Faulkner D (2001) Rogue waves – defining their characteristics for marine design. In: Olagnon M, Athanassoulis GA (eds) Rogue Waves 2000. Ifremer, France, pp 3–18

Forristall GZ (2005) Understanding rogue waves: Are new physics really necessary? In: Proc. 14th Aha Huliko'a Winter Workshop, Honolulu, Hawaii, 2005

Hasselman K, Hasselman S, Bartel K (1990) Use of a wave model as a validation tool for ERS-1 AMI wave products and as a input for the ERS-1 wind retreival algoritms. In: Max Planck Institut für Meteorologie, Hamburg, Report 55

Haver S (2005) A possible freak wave event measured at the Draupner jacket January 1 1995. In: Olagnon M, Prevosto M (eds) Rogue Waves 2004, Ifremer, France

Haver S, Andersen OJ (2000) Freak waves – rare realizations of a typical extreme wave population or typical realizations of a rare extreme wave population? In: Proc. 10th Int Offshore and Polar Eng Conf ISOPE, Seattle, USA, 2000, pp 123–130

Heavy Seas (2002) http://tv-antenna.com/heavy-seas/. Accessed 14 March 2008

Kharif C, Pelinovsky E (2003) Physical mechanisms of the rogue wave phenomenon. Eur J Mech/B – Fluid 22:603–634

Komen GJ, Cavaleri L, Donelan M et al (1994) Dynamics and modelling of ocean waves. Cambridge Univestity Press, Cambridge

Kurkin AA, Pelinovsky EN (2004) Freak waves: facts, theory and modelling. NNSTU, Nizhny Novgorod (In Russian)

Lavrenov IV (2003) Wind waves in ocean: dynamics and numerical simulations. Springer-Verlag, Heidelberg

Lawton G (2001) Monsters of the deep (The Perfect Wave). New Scientist 170 No 2297:28–32

Lechuga A (2006) Were freak waves involved in the sinking of the tanker "Prestige"? Nat. Hazards Earth Syst. Sci. 6: 973–978

Lehner SH (2005) Extreme wave statistics from radar data sets. In: Proc. 14th Aha Huliko`a Winter Workshop, Honolulu, Hawaii, 2005

Liu PC (2007) Freaque waves. http://freaquewaves.blogspot.com/2006_07_01_archive.html. Accessed 14 March 2008

Liu PC, MacHutchon KR (2006) Are there different kinds of rogue waves? In: Proc 25th Int Conf OMAE 2006, Hamburg, Germany, 2006, OMAE2006-92619:1–6

Lopatoukhin L, Boukhanovsky A, Divinsky B, Rozhkov V (2003) About freak waves in the oceans and seas. Proc of Russian Register of Shipping 26:65–73. (In Russian)

Mallory JK (1974) Abnormal waves on the south-east of South Africa. Inst Hydrog Rev 51:89–129

Melville WK, Romero L, Kleiss JM (2005) Extreme wave events in the Gulf of Tehuantepec. In: Proc 14th Aha Huliko`a Winter Workshop, Honolulu, Hawaii, 2005

Moreau F, translated by Olagnon M, Chase GA (2005) The Glorious Three. In: Olagnon M, Prevosto M (eds) Rogue Waves 2004. Ifremer, France. http://www.ifremer.fr/web-com/stw2004/rw/fullpapers/glorious.pdf. Accessed 14 March 2008

Mori N, Liu PC, Yasuda T (2002) Analysis of freak wave measurements in the Sea of Japan. Ocean Eng 29:1399–1414

Olagnon M (2000) Vagues extrêmes - Vagues scélérates. http://www.ifremer.fr/web-com/molagnon/jpo2000/. Accessed 13 March 2008

Olagnon M, Athanassoulis GA (eds) (2001) Rogue Waves 2000. Ifremer, France

Olagnon M, Magnusson AK (2004) Sensitivity study of sea state parameters in correlation to extreme wave occurrences. In: Proc. 14th Int Offshore and Polar Eng Conf ISOPE, Toulon, France, 2004, pp 18–25

Palmer AR (2002) A Rogue Wave. http://www.biology.ualberta.ca/courses.hp/biol361/WavePics/WavePics.htm. Accessed 14 March 2008

Paprota M, Przewlócki J, Sulisz W, Swerpel BE (2003) Extreme waves and wave events in the Baltic Sea. In: Rogue Waves: Forecast and Impact on Marine Structures. GKSS Research Center, Geesthacht, Germany

Peregrine DH, Bredmose H, Bullock G et al (2005) Violent water wave impact on a wall. In: Proc. 14th Aha Huliko`a Winter Workshop, Honolulu, Hawaii, 2005

Pinho de UF, Liu PC, Ribeiro CEP (2004) Freak waves at Campos Basin, Brazil. Geofizika 21:53–67

Pugh DT (1987) Tides, surges, and mean sea level. John Wiley & Sons, New York

References

Rabinovich AB (1993) Long ocean gravity waves: trapping, resonance, and leaking. Gidrometeoizdat, St. Petersburg (In Russian)

Rosenthal W (2005) Results of the MAXWAVE project. In: Proc. 14th Aha Huliko`a Winter Workshop, Honolulu, Hawaii, 2005. http://www.soest.hawaii.edu/PubServices/2005pdfs/Rosenthal.pdf. Accessed 14 March 2008

Stansell P (2004) Distributions of freak wave heights measured in the North Sea. Appl Ocean Res 26:35–48

Stansell P (2005) Distributions of extreme wave, crest and trough heights measured in the North Sea. Ocean Eng 32:1015–1036

Time (1954) Biggest tanker. Monday, 22 February, 1954. http://www.time.com/time/printout/0,8816,860511,00.html. Accessed 14 March 2008

Toffoli A, Lefevre JM, Bitner-Gregersen E, Monbaliu J (2005) Towards the identification of warning criteria: Analysis of a ship accident database. Appl Ocean Res 27:281–291

Torum A, Gudmestad OT (eds) (1990) Water Wave Kinematics. Kluwer, Dordrecht

Valenzuela GR (1978) Theories for the interaction of electromagnetic and ocean waves – A review. Bound Layer Meteorol 13:61–85

Wright JW (1968) A new model for sea clutter. IEEE Trans on Antennas and Propagation AP-16:217–223

Chapter 2
Deterministic and Statistical Approaches for Studying Rogue Waves

Depending on the objective in mind, two main approaches can be used for the water wave description, based on deterministic or statistical methods. Deterministic equations are very useful and powerful in understanding and describing the underlying physics of water waves; namely, they may be used in practice to estimate in detail wave impact upon structures and ships. Statistical equations are usually used to estimate typical wave motion and probability of this or that wave situation. When the sea surface elevation is such a complicated function of space and time, a statistical description is easier than a detailed description, but still may provide sufficient information about the waves.

In this chapter, we introduce first the basic equations governing the dynamics of water waves. The scales of the wavelength considered are long enough to neglect surface tension. Hence, the waves are called gravity waves since their main restoring force is gravity. Within the framework of water waves, we discuss and justify the different assumptions used to derive from the most complete system, the Navier-Stokes equations—a simplified set of equations describing realistic wave dynamics. In this way, the assumptions of incompressible and perfect fluid and irrotational motion are introduced successively to derive the simplified model. The simplified equations fall within the scope of the potential theory. Nevertheless, some of these assumptions may become questionable—for instance,, in shallow water where bottom friction can be important. Near the bottom a boundary layer of thickness of $O(2\nu/\Omega)$ develops, where ν and Ω are the molecular viscosity and the free surface wave frequency. So, for swells of 10 s, the boundary layer thickness is 0.17 cm with $\nu = 0.01\,\text{cm}^2\,\text{s}^{-1}$. The role of molecular viscosity in the formation of rogue waves can be considered as negligible. For turbulent boundary layers, the turbulent viscosity is much larger than the molecular viscosity ν and bottom friction may influence rogue wave dynamics. This aspect is discussed in Sect. 4.1.2. In the presence of breaking waves, the motion cannot be considered as irrotational and the dissipation of the waves is mainly due to turbulence (and not to molecular viscosity). Section 2.3 introduces concepts that will be used in subsequent chapters. Therefore, we focus attention on various physical mechanisms that contribute to the formation of extreme water wave events. Despite the complexity of the sea surface, we are aimed at describing quite simple realistic models that capture the essential features of rogue-wave phenomena.

2.1 Deterministic Equations

2.1.1 Mass and Momentum Conservation Equations

An Eulerian description of the fluid motion is adopted. The motion is described by the velocity field $\mathbf{U} = (U,V,W)^t$ as a function of time T, horizontal coordinates (X,Y) and vertical coordinate Z. The illustration of the problem geometry is provided in Fig. 2.1. The unperturbed surface coincides with the plane OXY at $Z = 0$, and the horizontal bed is situated at $Z = -D$. Typically, the waves are supposedly propagating along the OX direction.

The mass conservation or continuity equation is

$$\frac{\partial \rho}{\partial T} + \nabla \cdot (\rho \mathbf{U}) = 0, \tag{2.1}$$

or

$$\frac{D\rho}{DT} + \rho \nabla \cdot \mathbf{U} = 0, \tag{2.2}$$

where ρ is the water density, $\nabla \cdot$ is the divergence operator, and D/DT is the material derivative given by

$$\frac{D}{DT} = \frac{\partial}{\partial T} + (\mathbf{U} \cdot \nabla), \tag{2.3}$$

$\nabla = (\partial/\partial X, \partial/\partial Y, \partial/\partial Z)^t$ is the gradient operator and $(\bullet)^t$ indicates transposition.

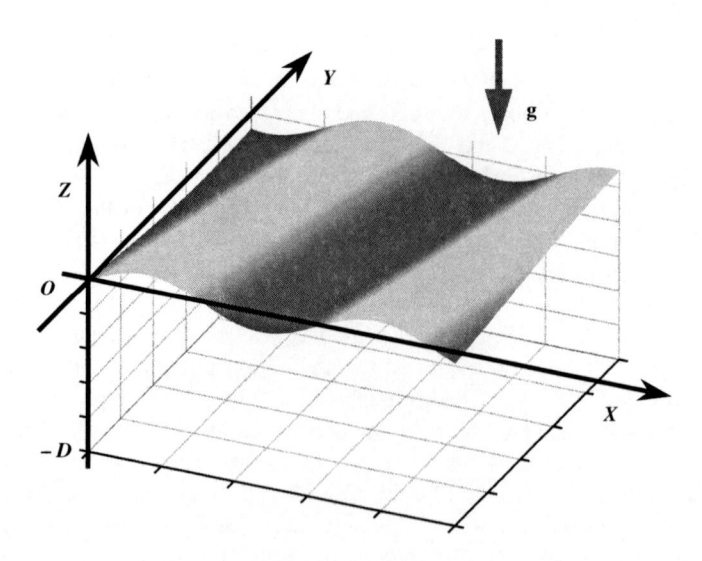

Fig. 2.1 Configuration of the problem

2.1 Deterministic Equations

The incompressibility condition of water reads $D\rho/DT = 0$, hence from the continuity equation we have

$$\nabla \cdot \mathbf{U} = 0. \tag{2.4}$$

The momentum-conservation equation, based on Newton's second law, reduces to the Navier-Stokes equation when considering water as an incompressible Newtonian fluid. The vector form of this equation is

$$\rho \frac{D\mathbf{U}}{DT} = -\nabla P + \rho \mathbf{F} + \mu \Delta \mathbf{U} \tag{2.5}$$

where P is the pressure, μ is the dynamic viscosity of the fluid, and Δ is the Laplacian operator $\Delta = \nabla \cdot \nabla$. The first and last terms on the Right Hand Side (RHS) of this equation correspond to pressure forces and viscous forces, respectively, while $\mathbf{F}$ is the body force due to the gravitational acceleration: $\mathbf{F} = \mathbf{g}$.

The corresponding $X-$, $Y-$ and $Z-$ momentum equations are given by

$$\rho \frac{DU}{DT} = -\frac{\partial P}{\partial X} + \rho F_X + \mu \Delta U, \tag{2.6}$$

$$\rho \frac{DV}{DT} = -\frac{\partial P}{\partial Y} + \rho F_Y + \mu \Delta V, \tag{2.7}$$

$$\rho \frac{DW}{DT} = -\frac{\partial P}{\partial Z} + \rho F_Z + \mu \Delta W, \tag{2.8}$$

where F_X, F_Y and F_Z are the components of the body forces $\mathbf{F}$ experienced by the fluid. Hence Eq. (2.5) is rewritten as follows:

$$\frac{D\mathbf{U}}{DT} = -\frac{1}{\rho}\nabla P + \mathbf{g} + \nu \Delta \mathbf{U}, \tag{2.9}$$

where $\nu = \mu/\rho$ is the kinematic viscosity.

Equation (2.9) may be written as follows:

$$\frac{\partial \mathbf{U}}{\partial T} + \frac{1}{2}\nabla(\mathbf{U}^2) = \mathbf{U} \times \omega - \frac{1}{\rho}\nabla P + \mathbf{g} + \nu \Delta \mathbf{U}, \tag{2.10}$$

where $\omega = \nabla \times \mathbf{U}$ is the vorticity. The operator $\nabla\times$ is the curl operator. By taking the curl of Eq. (2.9) and using Eq. (2.4), we obtain the vorticity equation

$$\frac{D\omega}{DT} = (\omega \cdot \nabla)\mathbf{U} + \nu \Delta \omega. \tag{2.11}$$

For 3D motions, the nonlinear term on the RHS of Eq. (2.11) is responsible for the vortex stretching and tilting while the linear term corresponds to the diffusion of vorticity due to viscosity.

Generally, water is considered as a weakly viscous fluid. In the vicinity of free surfaces and solid boundaries (the sea bottom), the thickness of the vortical layer is $O(\nu^{1/2})$. Hence, it will be assumed that the vortical part of the flow is confined to a thin boundary layer of thickness that is small compared to the other scales of the problem, so viscous effects are dropped from the equations.

We can consider that water waves have been generated from a fluid that was initially at rest—that is, from an irrotational motion. When the fluid is incompressible and inviscid, and external forces derive from a potential, the Kelvin-Lagrange theorem states that the motion remains irrotational. Hence, $|\omega| = 0$ and the velocity $\mathbf{U}$ derives from a potential $\phi(X,Y,Z,T)$ such that

$$\mathbf{U} = \nabla \phi. \tag{2.12}$$

Under the hypotheses of irrotational motion and inviscid fluid, Eqs. (2.4) and (2.10) become, respectively

$$\Delta \phi = 0 \tag{2.13}$$

and

$$\frac{\partial \mathbf{U}}{\partial T} + \frac{1}{2} \nabla \left(\mathbf{U}^2 \right) = -\frac{1}{\rho} \nabla P + \mathbf{g}. \tag{2.14}$$

Substituting $\nabla \phi$ for $\mathbf{U}$ in Eq. (2.14) gives

$$\nabla \left(\frac{\partial \phi}{\partial T} + \frac{1}{2} \nabla \phi \cdot \nabla \phi + \frac{P}{\rho} \right) - \mathbf{g} = 0 \tag{2.15}$$

Noting that $\mathbf{g} = (0,0,-g)^t$ so that $\mathbf{g} = \nabla(-gZ)$, the previous equation takes the following form

$$\nabla \left(\frac{\partial \phi}{\partial T} + \frac{1}{2} \nabla \phi \cdot \nabla \phi + \frac{P}{\rho} + gZ \right) = 0. \tag{2.16}$$

Integration with respect to space variables yields the Bernoulli equation

$$\frac{\partial \phi}{\partial T} + \frac{1}{2} \nabla \phi \cdot \nabla \phi + \frac{P}{\rho} + gZ = C(T). \tag{2.17}$$

The time dependent function $C(T)$ can be incorporated into the potential ϕ by the following transformation

$$\phi \to \phi + \int_0^T C(\xi)\,d\xi. \tag{2.18}$$

Thus, Eq. (2.17) is rewritten as follows:

$$\frac{\partial \phi}{\partial T} + \frac{1}{2} \nabla \phi \cdot \nabla \phi + \frac{P}{\rho} + gZ = 0. \tag{2.19}$$

To solve the Laplace equation (2.13), conditions on boundaries are needed.

2.1.2 Boundary Conditions

The fluid domain that is considered is bounded by two kinds of boundaries: the interface, which separates the air from the water; and the wetted surface of an impenetrable solid (the sea bottom, for instance). The air-sea interface is assumed to be a free surface whose equation is given by

$$S(X,Y,Z,T) = 0. \tag{2.20}$$

The kinematic boundary condition states that the normal velocity of the surface is equal to the normal velocity of the fluid at the surface. The normal velocity of the surface is

$$V_n = -\frac{1}{|\nabla S|}\frac{\partial S}{\partial T}, \tag{2.21}$$

and the normal velocity of the fluid is

$$U_n = \mathbf{n} \cdot \mathbf{U}. \tag{2.22}$$

where $\mathbf{n} = \nabla S/|\nabla S|$ is the unit vector normal to the surface.

The mathematical expression of the kinematic boundary condition is therefore

$$V_n = U_n. \tag{2.23}$$

Hence,

$$\frac{\partial S}{\partial T} + \mathbf{U} \cdot \nabla S = 0, \tag{2.24}$$

$$\frac{DS}{DT} = 0. \tag{2.25}$$

Equation (2.25) means that a fluid particle located on the free surface will remain on it.

An alternative form of the surface equation is

$$S(X,Y,Z,T) = \eta(X,Y,T) - Z = 0, \tag{2.26}$$

where $\eta(X,Y,T)$ represents the free surface elevation measured from $Z = 0$. Thus, Eq. (2.25) takes the form

$$\frac{\partial \eta}{\partial T} + U\frac{\partial \eta}{\partial X} + V\frac{\partial \eta}{\partial Y} - W = 0 \quad \text{on} \quad Z = \eta \tag{2.27}$$

or, equivalently

$$\frac{\partial \eta}{\partial T} + \frac{\partial \phi}{\partial X}\frac{\partial \eta}{\partial X} + \frac{\partial \phi}{\partial Y}\frac{\partial \eta}{\partial Y} - \frac{\partial \phi}{\partial Z} = 0 \quad \text{on} \quad Z = \eta. \tag{2.28}$$

Equations (2.23), (2.25) and (2.28) correspond to different forms of the kinematic boundary condition.

Since η and ϕ are both unknown on the free surface, a second boundary condition is needed: the dynamic boundary condition. This condition is derived from the Bernoulli equation (2.19). When surface tension is neglected, the pressure P in the fluid on the free surface is equal to the atmospheric pressure P_a. Hence, the Bernoulli equation (2.19) on the free surface takes the form

$$\frac{\partial \phi}{\partial T} + \frac{1}{2}\nabla\phi \cdot \nabla\phi + \frac{P_a}{\rho} + gZ = 0 \quad \text{on} \quad Z = \eta. \tag{2.29}$$

The atmospheric pressure P_a is chosen as reference and we can set P_a equal to zero without loss of generality. Hence,

$$\frac{\partial \phi}{\partial T} + \frac{1}{2}\nabla\phi \cdot \nabla\phi + gZ = 0 \quad \text{on} \quad Z = \eta. \tag{2.30}$$

For the rigid boundary, we have $S(X,Y,Z) = Z + D(X,Y) - 0$, thus $Z = -D(X,Y)$ is the equation of the sea bottom and Eq. (2.28) takes the form

$$\frac{\partial \phi}{\partial X}\frac{\partial D}{\partial X} + \frac{\partial \phi}{\partial Y}\frac{\partial D}{\partial Y} + \frac{\partial \phi}{\partial Z} = 0 \quad \text{on} \quad Z = -D(X,Y). \tag{2.31}$$

Although the Laplace equation is a linear partial differential equation, the difficulty in solving water wave problems arises from the nonlinearity of kinematic and dynamic boundary conditions. Furthermore, these equations apply on a surface that is unknown *a priori*. To summarize, the water wave problem reduces to solve the system of equations consisting of the Laplace equation (2.13), kinematic boundary condition (2.28), dynamic boundary condition (2.30) and sea bottom condition (2.31), with initial and boundary values for ϕ and η.

2.1.3 Linearization: Equations for Small Amplitude Waves

As emphasized in the previous section, we need values of the partial derivatives of the potential ϕ on a surface η that is unknown *a priori*. To solve the water wave equations, a free surface known *a priori* will be introduced through the linearization of the problem, which corresponds to an approximation of the nonlinear problem.

The nonlinearity of Eq. (2.30) is due to the presence of the convective term of the momentum equation, namely $(\mathbf{U} \cdot \nabla)\mathbf{U}$. Let us consider the simple example of a two-dimensional (2D) fluid motion. For waves propagating along the X direction, we consider the X-momentum equation and thus the corresponding nonlinear term is $U\partial U/\partial X + V\partial U/\partial Y$. Let us compare the first term to the linear term $\partial U/\partial T$ of the momentum equation. Let A, T_p and λ be the characteristic amplitude, period and wavelength of waves on the free surface, respectively. During a specific period, the fluid particles suffer displacements of order A. The corresponding fluid velocity and horizontal velocity gradient are then approximately A/T_p and $A/T_p\lambda$. Hence,

2.1 Deterministic Equations

$$U\frac{\partial U}{\partial X} = O\left(\frac{A^2}{\lambda T_p^2}\right)$$

and

$$\frac{\partial U}{\partial T} = O\left(\frac{A}{T_p^2}\right).$$

The linearization condition can, therefore, be written as

$$\left|U\frac{\partial U}{\partial X}\right| \ll \left|\frac{\partial U}{\partial T}\right| \rightarrow A \ll \lambda.$$

The condition for linearization of the equations is that the amplitude is small against the wavelength. Using $\lambda = 2\pi/K$, where K is the wavenumber, the previous equation yields to the condition

$$\varepsilon = AK \ll 1, \qquad (2.32)$$

where ε is the linearization parameter. Physically, this parameter is the wave steepness.

The water wave equations, which are nonlinear, can be transformed into a sequence of linear problems by using a perturbation procedure. Let us assume the following perturbation expansions in the parameter ε for the unknowns ϕ and η (i.e., Mei 1983 or Johnson 1997)

$$\phi(X,Y,Z,T) = \sum_{n=1}^{\infty} \varepsilon^n \phi_n(X,Y,Z,T), \qquad (2.33)$$

$$\eta(X,Y,T) = \sum_{n=1}^{\infty} \varepsilon^n \eta_n(X,Y,T). \qquad (2.34)$$

The temporal and spatial derivatives of the velocity potential ϕ, which occur in the free surface conditions (2.28) and (2.30), are expanded in the Taylor series about the still water level $Z = 0$:

$$\frac{\partial \phi}{\partial r}(X,Y,Z=\eta,T) = \sum \frac{\eta^n}{n!}\frac{\partial^n}{\partial Z^n}\left(\frac{\partial \phi}{\partial r}\right)(X,Y,Z=0,T), \qquad (2.35)$$

where r may represent temporal or spatial variables.

Substituting expansions (2.33), (2.34) and (2.35) into Eqs. (2.13), (2.28), (2.30), (2.31), and collecting the coefficients of the first power of ε, one finds

$$\Delta\phi_1 = 0, \quad -D < Z < 0, \qquad (2.36)$$

$$\frac{\partial \eta_1}{\partial T} - \frac{\partial \phi_1}{\partial Z} = 0 \quad \text{on} \quad Z = 0, \qquad (2.37)$$

$$\frac{\partial \phi_1}{\partial T} + g\eta_1 = 0 \quad \text{on} \quad Z = 0, \tag{2.38}$$

$$\frac{\partial \phi_1}{\partial X}\frac{\partial D}{\partial X} + \frac{\partial \phi_1}{\partial Y}\frac{\partial D}{\partial Y} + \frac{\partial \phi_1}{\partial Z} = 0 \quad \text{on} \quad Z = -D. \tag{2.39}$$

For small amplitude water waves $\varepsilon \ll 1$, we can ignore the terms of order $O(\varepsilon^n)$ with $n > 1$ in the expansions of (2.33), (2.34). Hence, the velocity potential and free surface elevation are approximated as

$$\phi(X,Y,Z,T) = \varepsilon \phi_1, \tag{2.40}$$

$$\eta(X,Y,T) = \varepsilon \eta_1. \tag{2.41}$$

The corresponding linear system of equations to be solved is

$$\Delta \phi = 0, -D < Z < 0, \tag{2.42}$$

$$\frac{\partial \eta}{\partial T} - \frac{\partial \phi}{\partial Z} = 0 \quad \text{on} \quad Z = 0, \tag{2.43}$$

$$\frac{\partial \phi}{\partial T} + g\eta = 0 \quad \text{on} \quad Z = 0, \tag{2.44}$$

$$\frac{\partial \phi}{\partial X}\frac{\partial D}{\partial X} + \frac{\partial \phi}{\partial Y}\frac{\partial D}{\partial Y} + \frac{\partial \phi}{\partial Z} = 0 \quad \text{on} \quad Z = -D. \tag{2.45}$$

2.1.4 Dispersion Relation

For the sake of simplicity, the bottom elevation, D, is considered to be constant. Hence, Eq. (2.45) becomes:

$$\frac{\partial \phi}{\partial Z} = 0 \quad \text{on} \quad Z = -D. \tag{2.46}$$

We look for a 2D periodic solution of the linear system of Eqs. (2.42), (2.43), (2.44) and (2.46) that admits the following velocity potential:

$$\phi(X,Z,T) = B\cosh[K(Z+D)]\sin(KX - \Omega T), \tag{2.47}$$

where B is a constant and Ω and K are the cyclic frequency and wave number, respectively. This form automatically satisfies the Laplace equation (2.42) and the bottom condition (2.46). Substituting (2.47) into the dynamical condition (2.44), one obtains

$$\eta(X,T) = \frac{B\Omega}{g}\cosh(KD)\cos(KX - \Omega T). \tag{2.48}$$

Let

$$A = \frac{B\Omega}{g}\cosh(KD). \tag{2.49}$$

2.1 Deterministic Equations

Hence,
$$\eta(X,T) = A\cos(KX - \Omega T), \qquad (2.50)$$
and the potential can be rewritten as follows:
$$\phi(X,Z,T) = \frac{Ag}{\Omega} \frac{\cosh[K(Z+D)]}{\cosh(KD)} \sin(KX - \Omega T). \qquad (2.51)$$

The linear dispersion relation is obtained by stating that the velocity potential (2.51) and the free surface elevation (2.50) correspond to nontrivial solutions satisfying the kinematic boundary condition (2.43),
$$\Omega^2 = gK\tanh(KD). \qquad (2.52)$$

The frequency Ω is given as a function of K in Fig. 2.2. Equations (2.50) and (2.51) represent 2D gravity waves of permanent form propagating with a constant phase velocity on water of uniform depth.

Equation (2.52) links the frequency Ω to the wave number, K. The phase velocity is given by
$$C_{ph} = \frac{\Omega}{K} = \sqrt{\frac{g}{K}\tanh(KD)}. \qquad (2.53)$$

Since $C_{ph}'(K) \neq 0, \forall K \neq 0$, the gravity water waves are dispersive. This is an important property of water waves, which means that waves of different wave numbers propagate at different phase velocities. Nevertheless, a stronger condition introduced by Whitham (1974) to define dispersive waves is $\forall K : \Omega''(K) \neq 0$.

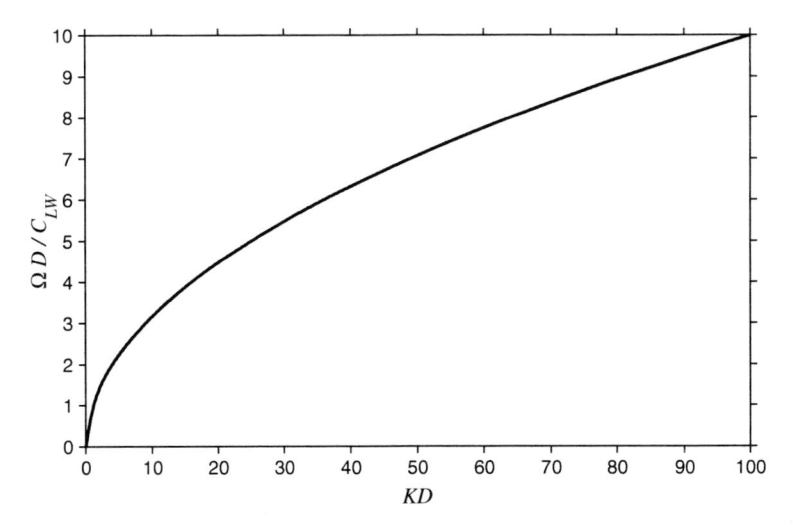

Fig. 2.2 Water wave dispersion relation curve as normalized frequency versus dimensionless water depth. The long wave velocity is defined as $C_{LW} = (gD)^{1/2}$

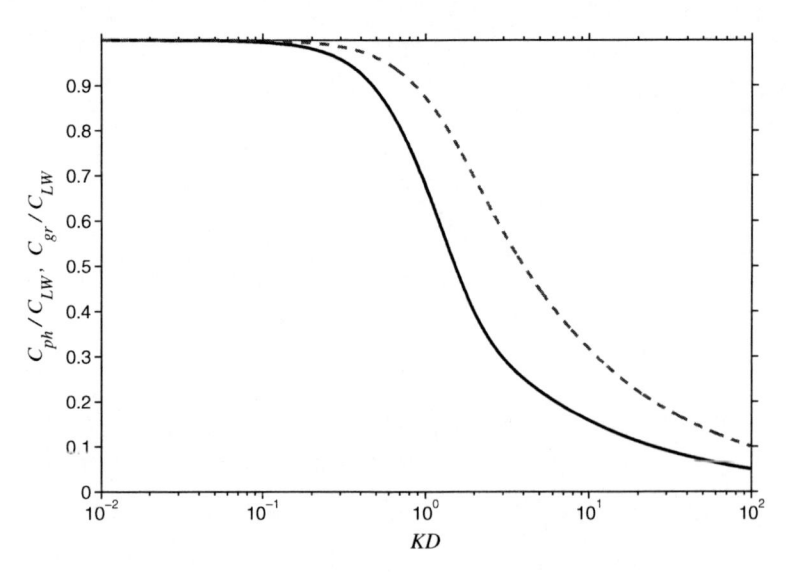

Fig. 2.3 Phase (*dotted line*) and group (*solid line*) velocity dependencies (C_{ph} and C_{gr} are normalized by C_{LW}) versus the dimensionless depth. Note logarithmic scale of the abscissa

The group velocity is defined as

$$C_{gr} = \frac{\partial \Omega}{\partial K} = \frac{g}{2\Omega}\left[\tanh(KD) + KD\left(1 - \tanh^2 KD\right)\right]. \quad (2.54)$$

In the shallow water limit $KD \to 0$, the group and phase velocities become equal and $C_{ph} \approx C_{gr} \to C_{LW}, C_{LW} = (gD)^{1/2}$; this means that the waves become nondispersive. The velocities C_{ph} and C_{gr} are given in Fig. 2.3 as functions of the dimensionless depth KD.

The 3D plane wave solution is given by the following formulas:

$$\eta(\mathbf{X}, T) = A\cos(\mathbf{K} \cdot \mathbf{X} - \Omega T) \quad (2.55)$$

and

$$\phi(\mathbf{X}, Z, T) = \frac{Ag}{\Omega}\frac{\cosh[|\mathbf{K}|(Z+D)]}{\cosh(|\mathbf{K}|D)}\sin(\mathbf{K} \cdot \mathbf{X} - \Omega T), \quad (2.56)$$

where $\mathbf{K}$ is the wave vector and $\mathbf{X} = (X,Y)^t$. The corresponding linear dispersion relation is

$$\Omega^2 = g|\mathbf{K}|\tanh(|\mathbf{K}|D). \quad (2.57)$$

Once the velocity potential ϕ is known, it is easy to calculate the velocity field $\mathbf{U} = \nabla\phi(\mathbf{X}, Z, T)$. The velocity components are

$$U = \frac{AgK_X}{\Omega}\frac{\cosh[|\mathbf{K}|(Z+D)]}{\cosh(|\mathbf{K}|D)}\cos(\mathbf{K} \cdot \mathbf{X} - \Omega T), \quad (2.58)$$

$$V = \frac{AgK_Y}{\Omega} \frac{\cosh\left[|\mathbf{K}|(Z+D)\right]}{\cosh(|\mathbf{K}|D)} \cos(\mathbf{K}\cdot\mathbf{X} - \Omega T), \tag{2.59}$$

$$W = \frac{Ag|\mathbf{K}|}{\Omega} \frac{\sinh\left[|\mathbf{K}|(Z+D)\right]}{\cosh(|\mathbf{K}|D)} \sin(\mathbf{K}\cdot\mathbf{X} - \Omega T), \tag{2.60}$$

where K_X and K_Y are the X– and Y– components of $\mathbf{K}$, respectively. The pressure $P(\mathbf{X};Z;T)$ is obtained from the Bernoulli equation (2.19).

For infinite depth $D \to \infty$, the bottom condition becomes

$$\nabla\phi \to 0 \text{ as } Z \to -\infty, \tag{2.61}$$

and the corresponding 2D gravity waves of permanent form propagating with a constant phase velocity are given by Eq. (2.50) and

$$\phi(X,Z,T) = \frac{Ag}{\Omega} \exp(KZ) \sin(KX - \Omega T) \tag{2.62}$$

with

$$\Omega^2 = gK. \tag{2.63}$$

2.2 Statistical Description

The second approach to studying waves is statistical. Water waves, of course, obey physical laws. They all, in principle, may be taken into account in a deterministic model, and therefore this model will be able (theoretically) to describe wave dynamics. In practice, this approach fails due to incomplete information about the initial state of the fluid, complexity of the physics, and growing fluctuations (this means that small perturbations with time may result in very different dynamics). Generally, the system of equations suffers from sensitive dependence on initial conditions. This feature is met in chaotic and turbulent systems. We know from our everyday experience that sea waves behave irregularly and unpredictably in even rather short time scales, although they show some periodicity. So, the dynamic system *Ocean* manifests *random* wave dynamics. Therefore, at certain sea conditions (significant wave height, wave age, winds, currents, etc.), different *realizations* (concerning the wave elevation – they are functions $\eta(\mathbf{X}, T)$) of sea waves are equally possible and may be considered as the object of investigation. The collection of realizations $\{\eta_j(\mathbf{X}, T)\}$ (integer subscript j counts them) builds an *ensemble*. In that way, the sea surface at one moment of time T_0 in one point $\mathbf{X}_0$ is represented by random functions numbered by the index j: $\eta_j(\mathbf{X}_0, T_0)$ with some statistical properties. This approach is referred to as *stochastic* and is aimed at a statistical description of sea wave dynamics.

The ultimate goal here is to describe and foresee the dynamics of certain realizations on the basis of the dynamics of averaged statistical characteristics. This approach is currently the center of attention of both theorists and experts, especially in the fields of ocean and atmospheric research; it is widely used in ocean engeneering. To obtain the time-dependence of statistical properties, one may perform stochastic simulations—i.e., to use deterministic models to compute a number of randomly chosen realizations (Monte Carlo simulations). Thus, one takes the position that the simulation of a sufficiently large but finite number of realizations represents the evolution of the whole ensemble in a statistical sense. The other approach is to compose and study models for direct computation of the evolution of statistical wave parameters. This is aimed at the theories that deal with spectral kinetic equations.

2.2.1 The Rayleigh Probability

Let us consider the surface displacement $\eta(\mathbf{X}, T)$—a function of space and time. Its autocorrelation function is defined as

$$R(\mathbf{X}, T, \mathbf{r}, \tau) = E\left[\eta\left(\mathbf{X}, T\right) \cdot \eta\left(\mathbf{X} + \mathbf{r}, T + \tau\right)\right], \quad (2.64)$$

where $E[\cdot]$ denotes statistical averaging over the ensemble of realizations $\eta_j(\mathbf{X}, T)$:

$$E\left[\eta\left(\mathbf{X}, T\right) \cdot \eta\left(\mathbf{X} + \mathbf{r}, T + \tau\right)\right] = \lim_{N \to \infty} \frac{1}{N} \sum_{j=1}^{N} \eta_j(\mathbf{X}, T) \cdot \eta_j(\mathbf{X} + \mathbf{r}, T + \tau). \quad (2.65)$$

In practice, N is finite, but it should be sufficiently large to provide a good estimate of the limit (2.65). Averaging over an ensemble is convenient for reproducible laboratory experimental conditions, but not the real ocean, where waves do not repeat themselves. For natural observations, a long time series is split into many shorter samples—"realizations"—that are used for averaging. This approach needs the random process to be *stationary* (i.e., its statistical properties do not depend on time). If these two ways of averaging result in the same statistics, the process is called ergodic. Although it is impossible to prove the ergodicity property for water waves via direct natural experiments, it is commonly invoked for the study of waves on the sea surface.

The statistical stationarity and statistical homogeneity in space imply that the autocorrelation function does not depend on $\mathbf{X}$ and T: $R = R(\mathbf{r}, \tau)$.

Averaging (2.64) may be also rewritten in terms of the probability function as

$$R(\mathbf{r}, \tau) = \int_{-\infty}^{\infty} \eta_1 \eta_2 f(\eta_1, \eta_2, \mathbf{r}, \tau) d\eta_1 d\eta_2, \quad (2.66)$$

where f is the two-point probability density function defined as

2.2 Statistical Description

$$f(\eta_1, \eta_2, \mathbf{r}, \tau) = \frac{\partial^2 F(\eta_1, \eta_2, \mathbf{r}, \tau)}{\partial \eta_1 \partial \eta_2}, \quad (2.67)$$

and the distribution function F measures probability such that $\eta(\mathbf{X}, T)$ and $\eta(\mathbf{X} + \mathbf{r}, T + \tau)$ do not exceed η_1 and η_2, respectively.

$$F(\eta_1, \eta_2, \mathbf{r}, \tau) = P(\eta(\mathbf{X}, T) \leq \eta_1 | \eta(\mathbf{X} + \mathbf{r}, T + \tau) \leq \eta_2). \quad (2.68)$$

Functions F and f do not depend on $\mathbf{X}$ and T if the field is both statistically homogeneous in space and stationary.

The probability distribution function or probability density function defines the statistical properties of the random field. To simplify the analysis of the statistics, integral parameters are often used. The nth statistical moment is defined as

$$\mu_n = E[\eta^n] = \int_{-\infty}^{\infty} \eta^n f(\eta) d\eta, \quad (2.69)$$

where f is the probability density function for η. Due to the normalization of the probability density function,

$$\mu_0 = 1. \quad (2.70)$$

The centered moments are defined as

$$\mu_n^c = E[(\eta - \mu)^n] = \int_{-\infty}^{\infty} (\eta - \mu)^n f(\eta) d\eta, \quad \mu \equiv \mu_1. \quad (2.71)$$

Only few low-order statistical moments have specific names due to their great importance in statistics. The first statistical moment μ in this instance is the mean water level. The variance σ^2 is equal to the second central moment

$$\sigma^2 = \mu_2^c = E\left[(\eta - \mu)^2\right], \quad (2.72)$$

and σ is the standard deviation. The skewness γ and kurtosis κ are defined through

$$\gamma = \frac{\mu_3^c}{\sigma^3} \quad (2.73)$$

and

$$\kappa = \frac{\mu_4^c}{\sigma^4}. \quad (2.74)$$

The skewness is usually used to estimate the vertical asymmetry of the sea surface elevation, whereas the kurtosis corresponds to the peakedness of the distribution when compared with the normal distribution (see Massel 1996).

The Central Limit Theorem proves that a superposition

$$\eta = \sum_j \eta_j \quad (2.75)$$

of statistically independent[1] variables η_j with mean values μ_j and variances σ_j^2 results in the Gaussian probability density

$$f(\eta) = \frac{1}{\sqrt{2\pi}\sigma} \exp\left[-\frac{(\eta - \mu)^2}{2\sigma^2}\right] \quad (2.76)$$

with mean

$$\mu = \sum_j \mu_j \quad (2.77)$$

and variance

$$\sigma^2 = \sum_j \sigma_j^2. \quad (2.78)$$

For Gaussian statistics, the skewness and kurtosis are $\gamma = 0$ and $\kappa = 3$, respectively.

Linear superposition of random periodic waves

$$\eta(\mathbf{X}, T) = \sum_n A_n \cos(\mathbf{K}_n \mathbf{X} - \Omega_n T + \theta_n) \quad (2.79)$$

is a natural representation of sea waves. Here, amplitudes A_n obey some probability distribution, and frequencies Ω_n and wave vectors $\mathbf{K}_n$ are dependent according to the dispersion relation; the wave phases θ_n are supposed to be uniformly distributed on the interval $[0, 2\pi]$. In this approximation, the surface elevation is described by the Gaussian statistics (2.76).

Let us now consider the linear superposition (2.79) of statistically independent Gaussian processes with variances σ_n^2. In the narrow-band assumption, the field may be represented in the following form

$$\eta = |B| \cos(\mathbf{K}_c \mathbf{X} - \Omega_c T + \varphi) \quad (2.80)$$

where $B = |B| \exp(i\varphi)$ is a slowly varying function of $\mathbf{X}$ and T, and σ_n^2 is rapidly decaying when values $\mathbf{K}_n$ (or Ω_n) are not close to $\mathbf{K}_c$ (or Ω_c, respectively). In this limit, the distribution for the linear wave amplitude $|B|$ is described by the Rayleigh function (Massel 1996)

$$f(|B|) = \frac{|B|}{\sigma^2} \exp\left(-\frac{|B|^2}{2\sigma^2}\right). \quad (2.81)$$

In the limit of small bandwidth, the wave height is twice the envelope, $H = 2|B|$, and therefore

$$f(H) = \frac{H}{4\sigma^2} \exp\left(-\frac{H^2}{8\sigma^2}\right), \quad (2.82)$$

---

[1] Two random variables are statistically independent if their joint probability density function may be factorized: $f(x,y) = f_x(x) \cdot f_y(y)$.

2.2 Statistical Description

and the probability that the wave height exceeds the value H (the exceedance probability) is

$$P(H) = 1 - F(H) = \exp\left(-\frac{H^2}{8\sigma^2}\right). \tag{2.83}$$

In Chap. 1, we introduced the significant wave height, which is the mean value of one-third of the highest waves. According to this definition and formula (2.82), the significant wave height is defined as (Massel 1996)

$$H_s = \frac{3}{4} \int_{\sigma\sqrt{8\ln 3}}^{\infty} \frac{H^2}{4\sigma^2} \exp\left(-\frac{H^2}{8\sigma^2}\right) dH \approx 4.004\sigma. \tag{2.84}$$

Integral (2.84) may be expressed through the error function (see Massel 1996). Usually a simplified relation is used, $H_s = 4\sigma$. Hence, formula (2.83) may be written in the convenient form

$$P(H) \approx \exp\left(-2\frac{H^2}{H_s^2}\right) \tag{2.85}$$

that helps to easily estimate the probability of high waves. For instance, a freak wave ($H > 2H_s$) should appear once among about 3,000 waves. For a typical sea wave period of 10 s, this gives the estimation that one should meet a freak wave every 8–9 h. In a Gaussian sea, a wave exceeding three times the significant height may occur once in 20 years.

Study of rogue waves in the framework of Gaussian statistics is already a tricky task. But waves (especially extreme waves) in the real ocean are obviously non-Gaussian due to various reasons: dissipation including wave breaking, insufficiently narrow spectrum, and nonlinear effects. Because rogue waves are rare events, and the sea state is persistently changing, the statistical stationarity condition also breaks down.

Nonlinear effects contribute to bound corrections to the wave shape as well as to the interaction between different harmonics, so periodic waves in superposition (2.79) become correlated. Due to the nonlinearity, waves become asymmetric: the crests are sharper and higher, while the troughs are flatter and shallower. The approximate bound nonlinear corrections may be taken into account with the help of the perturbation technique. In the deep-water case, the second-order small steepness ($KA << 1$) approximation gives

$$\eta(X,T) = A\cos(KX - \Omega T + \theta) + \frac{1}{2}KA^2\cos[2(KX - \Omega T + \theta)]. \tag{2.86}$$

for a regular monochromatic (Stokes) wave.

Assuming that the linear wave amplitude preserves the Rayleigh distribution, Formula (2.85) can be used to estimate the probability exceedance for wave crests (η_{cr}) and troughs (η_{tr}) by

$$P(\eta_{cr} > \eta) = \exp\left(-\frac{8}{H_s^2}\frac{(\sqrt{1+2K\eta}-1)^2}{K^2}\right) \qquad (2.87)$$

and

$$P(\eta_{tr} > \eta) = \exp\left(-\frac{8}{H_s^2}\frac{(\sqrt{1-2K\eta}-1)^2}{K^2}\right). \qquad (2.88)$$

Formulas (2.87) and (2.88) predict that extreme waves have larger crests than troughs. We should note, however, that representation (2.86) does not lead to a change of the crest-to-trough wave height at this level of accuracy.

Different types of modified distribution functions, taking into account weak nonlinear bound corrections, were developed in Tayfun (1980), Tung and Huang (1985), and Mori and Yasuda (2002) and many others (see survey by Prevosto 2001); other modifications of the Rayleigh distribution are being developed, as are empirical formulas. Apparently, second-order statistical models turn out to be insufficient for the adequate description of rogue waves (Bitner-Gregersen and Magnusson 2005, Rosenthal 2005, Petrova et al. 2007). Nonlinear corrections of higher orders should be taken into account (Creamer et al. 1989, Huang et al. 1983, Zhang et al. 1999); these corrections may enhance the probability of high waves by ten (Prevosto and Bouffandeau 2002) or even one hundred (Stansell 2004, Forristall 2005) times! Since nonlinear properties of surface waves depend on depth, the depth is one more parameter in the statistical model (see Massel 1996).

The considered theory is applied to one-point observations. Real needs and recent 3D observations require development of a statistical model describing wave probability over a specific area. Reduction from the point statistics is not trivial for this purpose (Forristall 2005, Socquet-Juglard et al. 2005), and may be very important. Thus, Forristall (2005) estimates that for the air gap under a fixed structure with a deck $50\,\text{m} \times 50\,\text{m}$, the maximum wave crest is almost 20% higher than the one expected at a single point.

The sea state is rather changeable; this results in failure of the condition of statistical stationarity. Donelan and Magnusson (2005) and Müller et al. (2005) show how the probability of high waves grows in a mixed sea constituted by two wave trains. Baxevani and Rychlik (2006) considered a Gaussian sea evolving in time and also studied the effects of wave spreading. They report that the neglect of these effects leads to an underestimation of the high wave probability.

2.2.2 Wave Spectra

The Fourier transform of the autocorrelation function R gives the *wave spectrum*

$$\hat{S}(\mathbf{K},\Omega) = \frac{1}{(2\pi)^3}\int_{-\infty}^{\infty} R(\mathbf{r},\tau)\cdot\exp[i(\mathbf{Kr}-\Omega\tau)]d\mathbf{r}d\tau. \qquad (2.89)$$

2.2 Statistical Description

Here, $\mathbf{K} = (K_X, K_Y)$ is the wave vector and Ω is the frequency. The *frequency spectrum* and *wave vector spectrum* (or *two-dimensional wavenumber spectrum* or *spatial spectrum*) are defined, respectively, by

$$\hat{S}(\Omega) = \int_{-\infty}^{\infty} \hat{S}(\mathbf{K}, \Omega) \, d\mathbf{K}, \tag{2.90}$$

and

$$\hat{S}(\mathbf{K}) = \int_{-\infty}^{\infty} \hat{S}(\mathbf{K}, \Omega) \, d\Omega. \tag{2.91}$$

The wavenumber spectrum is defined as

$$\hat{S}(K) = \int_{-\pi}^{\pi} K \hat{S}(\mathbf{K}) \, d\alpha, \text{ where } \mathbf{K} = (K \cos \alpha, K \sin \alpha), \tag{2.92}$$

and $K = |\mathbf{K}| > 0$ is the wavenumber. The *directional spectrum* is

$$\hat{S}(\alpha) = \int_{0}^{\infty} dK \int_{-\infty}^{\infty} d\Omega K \hat{S}(\mathbf{K}, \Omega). \tag{2.93}$$

The frequency and wave vector (wavenumber) spectra can be related to one another; this can be achieved with the help of the dispersion relation. For instance, for the deep-water case, the dispersion relation is as follows

$$K dK = 2 \frac{\Omega^3}{g^2} d\Omega, \tag{2.94}$$

and hence

$$\hat{S}(\Omega) = \frac{2\Omega^3}{g^2} \hat{S}(K). \tag{2.95}$$

For a real process—statistically stationary and statistically homogeneous in space—the correlation function possesses the symmetry property $R(-\mathbf{r}, -\tau) = R(\mathbf{r}, \tau)$. Then the spectrum is real, and $\hat{S}(-\mathbf{K}, -\Omega) = \hat{S}(\mathbf{K}, \Omega)$. That is why only one half of the spectrum is commonly used in the analysis: $\Omega > 0$ for the frequency spectrum, and $K > 0$ for the wavenumber spectrum.

In the first approximation, the wave field may be represented as a linear superposition of periodic waves (2.79). To see this, let us consider a single cosine wave

$$\eta(\mathbf{X}, T) = A_0 \cos(\mathbf{K}_0 \mathbf{X} - \Omega_0 T + \theta), \tag{2.96}$$

where A_0, $\mathbf{K}_0$ and Ω_0 are defined, but θ is a random value uniformly distributed within the interval $[-\pi, \pi]$. Then, the corresponding correlation function is

$$R(\mathbf{r},\tau)$$
$$= \int_{-\pi}^{\pi} A_0 \cos(\mathbf{K}_0 \mathbf{X} - \Omega_0 T + \theta) A_0 \cos[\mathbf{K}_0(\mathbf{X}+\mathbf{r}) - \Omega_0(T+\tau) + \theta] \frac{d\theta}{2\pi} \quad (2.97)$$
$$= \frac{A_0^2}{2} \cos(\mathbf{K}_0 \mathbf{r} - \Omega_0 \tau).$$

Furthermore, the wave spectrum for a single cosine wave reads

$$\hat{S}(\mathbf{K},\Omega) = \frac{1}{(2\pi)^3} \int_{-\infty}^{\infty} R(\mathbf{r},\tau) \cdot \exp[i(\mathbf{K}\mathbf{r} - \Omega\tau)] d\mathbf{r} d\tau$$
$$= \frac{A_0^2}{4} \left(\delta(\mathbf{K}+\mathbf{K}_0) \delta(\Omega+\Omega_0) + \delta(\mathbf{K}-\mathbf{K}_0) \delta(\Omega-\Omega_0) \right) \quad (2.98)$$

Hence, for the linear superposition of periodic waves (2.79), the spectrum has the form

$$\hat{S}(\mathbf{K},\Omega) = \sum_n \frac{A_n^2}{4} \left(\delta(\mathbf{K}+\mathbf{K}_n) \delta(\Omega+\Omega_n) + \delta(\mathbf{K}-\mathbf{K}_n) \delta(\Omega-\Omega_n) \right). \quad (2.99)$$

Thus, the wave spectrum is represented by Dirac delta functions and has non-zero values in the $(\mathbf{K}, \Omega)$ space only at points corresponding to the waves represented in the superposition (2.79).

It is well known that the total energy of a linear plane progressive wave (2.96) is defined as

$$En = \rho g \frac{A_0^2}{2}. \quad (2.100)$$

Alternatively, Formula (2.98) gives

$$En = \rho g \int_{-\infty}^{\infty} \hat{S}(\mathbf{K},\Omega) d\mathbf{K} d\Omega. \quad (2.101)$$

Therefore, the wave spectrum has the meaning of the wave energy distribution in the space of wave vectors and frequencies; the quantity $\rho g \hat{S}$ is called the *energetic spectrum*.

The wave amplitudes may be expressed through the relation

$$A_n^2 = 2 \int \hat{S}(\mathbf{K}_n,\Omega_n) d\mathbf{K} d\Omega, \quad (2.102)$$

where the integration is effective only in closed intervals around $\pm \mathbf{K}_n$ and $\pm \Omega_n$. Relations (2.99) and (2.102) allow us to define the spectrum as the squared absolute value of the Fourier transform of the process. The relationship between the spectrum and the autocorrelation function (2.89) is then called the Wiener-Khintchine Theorem. The spectrum concept is a powerful tool for investigating time series, since it displays the distribution of wave energy among frequencies and scales represented by harmonics. Data processing in the spectral space (such as filtering) may

2.2 Statistical Description

be very powerful. Although the wave height, peak period, and main wave direction are sufficient to describe sea states for most practical purposes, Olagnon and Magnusson (2004) pointed out at the same time that it is likely that no spectral parameter alone can provide useful information on the risk and potentially abnormal wave events when it is estimated. Thus, this approach needs improvement to be applied to the needs of rogue wave research.

2.2.2.1 Frequency Spectrum

For most of the numerous measurements of sea waves represented by time series at one spatial point, only the frequency spectrum may be obtained. Since it is an even function of frequency, instead of the symmetric function $\hat{S}(\Omega), \Omega \in (\infty; \infty)$, only one part is used in experimental practice (so-called nonsymmetric spectrum): $S(\Omega) = 2\hat{S}(\Omega), \Omega \in [0; \infty)$. The longer the record is, the more statistical material it provides; on the other hand, sea conditions may change if the realization takes too long. Usually wave record samples of duration 10–30 min are retrieved for analysis to fulfill these contradictory requirements. The relationship between the spectrum and the wave amplitudes persists

$$A_n = \sqrt{2S(\Omega_n)\Delta\Omega}, \qquad (2.103)$$

where $\Delta\Omega$ is the frequency discretization interval. As an estimator for frequency spectrum $S(\Omega)$, the Fourier transform of the wave field is usually employed in practice:

$$S(\Omega) \cong 2|\eta_\Omega|^2, \quad \eta_\Omega = \frac{1}{T}\int_0^T \eta(t)\exp(i\Omega t)\,dt. \qquad (2.104)$$

When analyzing the wave spectrum, the spectral moments are frequently used; they are, in general, defined as

$$m_n = \int_0^\infty \Omega^n S(\Omega)d\Omega. \qquad (2.105)$$

Inversing the Fourier transform (2.89), one obtains

$$R(\tau = 0) = \int_{-\infty}^\infty \hat{S}(\Omega)\,d\Omega = \int_0^\infty S(\Omega)\,d\Omega, \qquad (2.106)$$

therefore the zero spectral moment is expressed through the second statistical moment

$$m_0 = \mu_2, \qquad (2.107)$$

or for the case of a field with zero mean,

$$m_0 = \mu_2^c = \sigma^2. \qquad (2.108)$$

The mean wave frequency and wave period are defined as

$$\Omega_p = \frac{m_1}{m_0}, \quad T_p = \frac{2\pi}{\Omega_p} = 2\pi \frac{m_1}{m_0}, \tag{2.109}$$

although other possible ways to define the mean frequency exist.

Central moments

$$m_n^c = \int_0^\infty (\Omega - \Omega_p)^n S(\Omega) d\Omega \tag{2.110}$$

are also used. The central moment m_2^c is a measure of the concentration of the spectral wave energy around the frequency Ω_p, which characterizes the spectral width through the dimensionless parameter

$$\delta_\Omega = \frac{1}{\Omega_p} \sqrt{\frac{m_2^c}{m_0}}. \tag{2.111}$$

The spectral shape displays the distribution of energy between scales and thus contains information about the physical mechanisms supporting and generating the waves. Concerning wind-generated waves, the energy growth due to the wind action is balanced by the wave interactions—which transfer energy between frequencies—and energy dissipation. Following the hypothesis of similarity for ocean waves, the energy spectrum should be represented by a function of the form (Massel 1996)

$$S(\Omega) = S(\Omega, X_f, T, U_w, g), \tag{2.112}$$

where X_f is the fetch, T is related to the wave age, and U_w is the wind velocity, or alternatively,

$$S(\Omega) = S(\Omega, g, \sigma, \Omega_p). \tag{2.113}$$

The suggested spectral shapes usually have the general form

$$S(\Omega) = C_1 \Omega^{-p} \exp\left(-C_2 \Omega^{-q}\right). \tag{2.114}$$

One of the most popular spectrums was suggested by Pierson and Moskowitz (1964) on the basis of theoretical discoveries and field data analysis:

$$\begin{aligned} S(\Omega) &= \alpha g^2 \Omega^{-5} \exp\left(-B \left(\frac{\Omega U_X}{g}\right)^{-4}\right), \\ &= \alpha g^2 \Omega^{-5} \exp\left(-\frac{5}{4} \left(\frac{\Omega}{\Omega_p}\right)^{-4}\right) \end{aligned} \tag{2.115}$$

where $\alpha = 8.1 \times 10^{-3}$, and $B = 0.74$. It was proposed for a fully developed sea, when the wave phase speed is equal to the wind speed. It is controlled by a single parameter, which is the wind speed.

2.2 Statistical Description

The Joint North Sea Wave Project (JONSWAP) Spectrum extends the Pierson-Moskowitz Formula (2.115) to include fetch-limited seas through inclusion of one more governing parameter manifesting peakedness, γ

$$S(\Omega) = \alpha g^2 \Omega^{-5} \exp\left(-\frac{5}{4}\left(\frac{\Omega}{\Omega_p}\right)^{-4}\right) \gamma^\delta \qquad (2.116)$$

$$\delta = \exp\left[-\frac{(\Omega - \Omega_p)^2}{2\sigma_0^2 \Omega_p^2}\right]$$

where $\gamma = 3.3$; $\sigma_0 = 0.07$, if $\Omega \leq \Omega_p$; and $\sigma_0 = 0.09$, if $\Omega > \Omega_p$.

$$\alpha = 0.076 \left(\frac{gX_f}{U_w^2}\right)^{-0.22}, \qquad (2.117)$$

$$\Omega_p = 7\pi \frac{g}{U_w}\left(\frac{gX_f}{U_w^2}\right)^{-0.33}. \qquad (2.118)$$

The JONSWAP Spectrum was built on the basis of an extensive wave measurement in the North Sea. This area is very popular in recent studies, owing to its great economical importance and large number of instrumental observations. With an extra free parameter, this shape is a convenient model spectrum. Other spectral shapes have been suggested and may be found in Massel (1996), but will not be considered in the present book.

2.2.3 Kinetic Models

So far, only statistically stationary and spatially homogeneous processes have been considered. This approach does not describe a realistic sea state where the wave field is evolving and changing from one area to another. The variability of waves may be computed when local (in time and space) statistics, that are represented by the wave spectrum, are considered. So, the spectrum function $\hat{S}$ depends on slow variables $\mathbf{X}$ and T, and all the wave conditions and statistical wave parameters may vary slowly in space and time.

Energy conservation results in the energy balance equation if there are no currents. Generally, it is given by the balance equation for the wave action, N (Whitham 1974)

$$\frac{dN}{dT} = G, \qquad (2.119)$$

where

$$N = \frac{\hat{S}}{\Omega_i}. \qquad (2.120)$$

Value Ω_i is the intrinsic frequency (in absence of current) that is related to the wavenumber through Eq. (2.57),

$$\Omega_i^2 = g|\mathbf{K}|\tanh(|\mathbf{K}|D), \tag{2.121}$$

while the observed (apparent) frequency in the presence of the current with velocity $\mathbf{U}_c$ is given by

$$\Omega = \Omega_i + \mathbf{K}\mathbf{U}_c. \tag{2.122}$$

The term G on the RHS of Eq. (2.119) defines the external action and losses, is called the *collision integral*, and takes into account different physical mechanisms: income of energy from the wind (pressure fluctuations, wave-flow linear and non-linear interactions); interaction with the atmosphere and sea turbulence; dissipation due to bottom friction; wave breaking; and spectral nonlinear exchange, etc.

$$G = \sum_n G_n \tag{2.123}$$

The wave action is a function of wave vector, apparent frequency, and slow variables $\mathbf{X}$ and T, thus the conservation of volume in space $(\mathbf{X}, \mathbf{K}, \Omega)$ results in

$$\frac{dN}{dT} = \frac{\partial N}{\partial T} + \frac{\partial N}{\partial \mathbf{X}}\frac{d\mathbf{X}}{dT} + \frac{\partial N}{\partial \mathbf{K}}\frac{d\mathbf{K}}{dT} + \frac{\partial N}{\partial \Omega}\frac{d\Omega}{dT} = G. \tag{2.124}$$

The assumption that all changes happen much slower (in time and space) than the period and length of the waves allows us to use the ray theory. Hence the wave filed may be represented as

$$\eta = A(\mathbf{X}, T)\exp(i\theta), \tag{2.125}$$

where θ is the phase. It is natural to define the local wave vector and frequency as

$$\mathbf{K}(\mathbf{X}, T) = \frac{\partial \theta}{\partial \mathbf{X}}, \text{ and } \Omega(\mathbf{K}, \mathbf{X}, T) = -\frac{\partial \theta}{\partial T}. \tag{2.126}$$

These relations then give the kinematic conservation equations

$$\frac{\partial \mathbf{K}}{\partial T} + \nabla \Omega = 0, \quad \frac{\partial K_X}{\partial Y} = \frac{\partial K_Y}{\partial X}, \tag{2.127}$$

where $\mathbf{K} = (K_X, K_Y)$, $\nabla = (\partial/\partial X, \partial/\partial Y)$. The second equation means that the wave vector field is irrotational. It follows then that

$$\frac{d\mathbf{K}}{dT} = -\frac{\partial \Omega}{\partial \mathbf{X}}, \quad \frac{d\mathbf{X}}{dT} = \frac{\partial \Omega}{\partial \mathbf{K}}, \text{ and } \frac{d\Omega}{dT} = \frac{\partial \Omega}{\partial T}, \tag{2.128}$$

which means that the quantity Ω may be understood as the Hamiltonian, while $\mathbf{X}$ is the position and $\mathbf{K}$ is the momentum. With the help of relations (2.128), the balance equation (2.124) transforms into

$$\frac{\partial N}{\partial T} + \frac{\partial N}{\partial \mathbf{X}}\frac{\partial \Omega}{\partial \mathbf{K}} - \frac{\partial N}{\partial \mathbf{K}}\frac{\partial \Omega}{\partial \mathbf{X}} + \frac{\partial N}{\partial \Omega}\frac{\partial \Omega}{\partial T} = G. \tag{2.129}$$

2.2 Statistical Description

The spectral energy balance equation (2.129) is called *the radiative transfer equation*, or *the transport equation*, or *the kinetic equation* and is used for forecasting spectral changes of sea waves. The first step for describing the evolution of the wave spectrum was done by Gelci et al. (1956, 1957) who introduced the concept of the spectral transport equation.

The first term at the Left Hand Side (LHS) of (2.129) expresses the local time evolution of the spectrum, and the second one represents the evolution of the spectrum for the horizontally inhomogeneous wave field and provides the energy transport with the group velocity $\partial\Omega/\partial\mathbf{K}$. The third term in (2.129) reflects the effects of refraction and shoaling due to the spatial change of the dispersion relation (because of variable bathymetry or currents), while the fourth term describes the temporal evolution of the dispersion relation due to changing conditions.

The basic difficulty in solving Eq. (2.129) is an evaluation of the source function G. The theory of weak nonlinear interactions for wind-induced waves was first formulated by Hasselmann (1962, 1968); it involves nine terms in the sum (2.123). The terms representing the wave-wave interactions are quite bulky and make Eq. (2.129) an integro-differential type. The nonlinear interaction coefficients have been obtained through tedious computations for low-order nonlinear interactions (up to five-wave interactions) (see Zakharov 1974, 1999; Krasitskii 1994; and Davidan et al. 1985; Lavrenov 2003; Janssen 2004; Polnikov 2007). In this application, the Hamiltonian approach is very convenient when the theory is expressed in terms of specially defined canonical variables (Zakharov 1968, 1974, 1999; Zakharov and Kuznetsov 1997; Polnikov 2007). The kinetic equation may be obtained rigorously starting from the primitive hydrodynamic equations, or from weakly nonlinear dynamical models such as the Zakharov equations (Zakharov 1974, 1999; Krasitskii 1994). However, the still open question is whether—and, if so, how much and under which conditions—the numerical evolution of a spectrum evaluated with the kinetic equation corresponds to the spectrum obtained with the full integration of the dynamical model starting from the actual surface distribution (Cavaleri 2005).

It has already been pointed out that in addition to the bound corrections to the wave shapes, the wave nonlinearity results in interactions between Fourier harmonics (that are independent in the linear limit). Due to this interaction, the energy in the spectral space may focus on one scale (uniform waves), or spread over many frequencies, and under certain conditions form very steep intensive waves. On the surface of deep water, the main part of the wave-wave nonlinear interactions in G is represented by the four-wave interaction. The spectral energy balance equation was used by Janssen (2003) to find the corrections to the Gaussian statistics of high sea waves when four-wave interactions are taken into account. The results were compared against the stochastic Monte Carlo simulations of dynamical models. The nonlinear effects in wave dynamics causing significant wave enhancement will be considered further in Chaps. 4 and 5.

Both considered approaches—deterministic and statistical—have strong and weak points that indicate their successful application for different purposes. In Sect. 2.2.1, we describe the one-point approach imposed by the instrumental data that is presently available. Other restrictions are due to the hypotheses employed by the approach (such as statistical homogeneity and stationarity). Realistic statistical models are still developing, and they should be verified versus observations. Some statistical aspects of nonlinear waves over deep, shallow, and coastal waters are discussed further in Sects. 4.4, 4.7, 5.3, and 5.5.

Unlike field observations and laboratory experiments, which usually give either temporal data at a few locations in space or spatial data at a few instants, the direct numerical simulations may provide both temporal and spatial data of a large-scale wave field. At the same time, the complexity and nonreproducibility of sea wave dynamics make the phase-resolving, long-time dynamic simulations practically useless. Draper in 1964 remarked that it is probably not possible to predict rogue-wave occurrence at a given time and space, although their probability might be estimated exceeding the framework of the stationary Gaussian process.

To obtain realistic rogue statistics, wave data collection is probably not the most adequate approach. Besides the problem of instrumental measurement briefly discussed in Chap. 1, one will face the following question: is the observed extreme wave a very rare realization from the typical slightly non-Gaussian sea surface population, or is it a typical realization of a very rare and strongly non-Gaussian sea surface population (Haver and Andersen 2000, Haver 2005)? The ensemble technique widely used in meteorology is promising. Each simulation is obtained by perturbing the conditions and/or initial sea wave field and letting the system evolve. Given the spectrum at a certain instant of time and location, one can choose a possible realization or a number of realizations. This would provide robust statistics of the sea surface, inclusive of all the nonlinear processes. Rather than acting only on the phases, one could act on the spectrum, both as amplitude and directions. In addition, the perturbations could be done not at random, but acting, for example, on specific groups of components chosen according to the situation (Cavaleri 2005).

2.3 Possible Physical Mechanisms of Rogue Wave Generation

Freak waves have been observed in basins of arbitrary depth (in deep as well as shallow water) with or without current and with or without wind. To resume, they may potentially occur everywhere on the ocean surface under any sea state conditions (see Chap. 1).

Before briefly presenting the main physical mechanisms leading to huge waves, let us introduce the critical depth parameter that allows separation between deep water and shallow water. Wave properties depend crucially on the water depth. This evident feature follows from the dispersion relation. While very long waves are not dispersive, dispersion becomes essential for shorter waves (see Figs. 2.2, 2.3). Furthermore, the dependence of the dispersion on water depth results in different manifestations of nonlinear wave-wave interactions. In shallow water, three-wave interactions play a major role, whereas in deep or moderately deep water, the main contribution to nonlinear wave interactions comes generally from four-wave interactions. The variety of nonlinear properties of sea waves over finite depths provides a rich and complex picture of nonlinear instabilities that could spawn rogue waves.

The natural parameter used to define deep water or shallow water conditions is the dimensionless depth, KD, where K is the wave number and D is the water depth. Following Fenton (1979), shallow water conditions correspond to $KD < \pi/4$, otherwise waves are propagating on finite depth or deep water. For $KD > \pi/4$, the Stokes-like expansion is relevant to calculating accurately nonlinear wave fields, whereas in shallow water it is the cnoidal-like expansion that prevails. In deep water and finite depth, the small parameter used in the Stokes expansion is the wave steepness AK (A denotes the wave amplitude), while for shallower water this parameter becomes A/D. In this book, we use the critical value $\pi/4$ of the normalized depth KD to separate deep water from shallow water. Note that in Chap. 3, the distinction between deep and shallow water is not used, whereas Chaps. 4 and 5 consider physical mechanisms that act in deep and shallow seas, respectively.

As it was noted previously, in addition to the dispersive parameter KD, there exist nonlinear parameters AK and A/D for deep water and shallow water, respectively. The parameter AK was already introduced in Sect. 2.1.3 for linearization of the equations. Chapter 3 will focus on linear aspects of rogue occurrence, while Chaps. 4 and 5 will consider nonlinear aspects. The main efforts are focused on the nonlinear and strongly nonlinear dynamics of the rogue-wave phenomenon based on recent research, because such waves are more dangerous. We also collect the results of statistical processing of natural registrations in Sect. 4.7.2; they, in part, support theoretical trends, although in the present state they are, in fact, often contradictory.

There are various physical mechanisms generating rogue waves on the sea surface. They can be due to geometrical focusing of directionally spread waves, refraction phenomena (presence of variable current or bottom topography), frequency modulation (dispersive focusing or modulational instability of Benjamin-Feir type), or soliton interactions that may produce wave energy that focuses in a small area. It was recently suggested that wave fields resulting from the nonlinear interaction of two wave systems (crossing seas) could be unstable to modulational instability and therefore produce rogue-wave occurrence (see Sect. 4.6). In this section, the different mechanisms are briefly listed and presented, they will be investigated and discussed deeply in the subsequent chapters. These effects have been previously reviewed by Kharif and Pelinovsky (2003) and Dysthe et al. (2008).

2.3.1 Wave-Current Interaction

Freak-wave occurrence on currents is a well-understood problem (see Sect. 3.4) that can explain the formation of rogue waves when wind waves or swells are propagating against a current. Besides more sophisticated models, the use of basic equations describing conservation of kinematical and dynamical properties of water-wave fields can be very convenient in determining the transformation of water waves by currents.

2.3.2 Geometrical or Spatial Focusing

Meanwhile, freak waves are seemingly observed throughout the world's oceans without significant currents as well. Underwater topography modifies the wave propagation. The result is spatial variations of the kinematic and dynamic variables of the problem that can be solved by the use of ray theory. Hence, rogue wave occurrence corresponds to caustics (see Sect. 3.1).

2.3.3 Focusing Due to Dispersion: The Spatio-Temporal Focusing

The spatio-temporal wave focusing due to the dispersive nature of water waves is a classic mechanism yielding wave-energy concentration in a small area (see Sect. 3.2). This effect, which can occur at the sea surface, can be reproduced easily in a laboratory experiment. Interactions with sea currents and wind flows represent specific features of sea waves. The effect of wind action is taken into account within the linear approximation in Sect. 3.3.

It is evident that once the wave steepness becomes finite, nonlinearity needs to be included. Effects of water wave nonlinearity on the above processes are discussed in Chap. 3 and further in Chaps. 4 and 5. Both weak and strong nonlinear approaches are presented. The achieved conclusions are verified against available results of laboratory experiments.

2.3.4 Focusing Due to Modulational Instability

This phenomenon is essentially nonlinear. Nonlinear uniform wave trains suffer an instability known as the Benjamin-Feir instability, which produces growing modulations of the envelope. These modulations that evolve into short groups of steep waves correspond to a nonlinear focusing of the wave energy. At the maximum of modulation, rogue waves can occur followed by the demodulation of the envelope. Rogue waves resulting from the modulational instability are considered in Chap. 4.

2.3.5 Soliton Collision

Uniform wave trains under modulational instability transform into a system of envelope solitons that may collide to give rise to huge wave events. Instability of quasi-solitons of large amplitude followed by collapse has been suggested as a proper scenario of rogue wave occurrence as well. These mechanisms that are working on finite and infinite water depths are presented in Chap. 4. Rogue waves can also occur in shallow water due to soliton interaction. The latter aspect is discussed in Chap. 5.

List of Notations

C_{gr}	group velocity
C_{LW}	long wave velocity
C_{ph}	phase velocity
D	water depth
D/DT	material derivative
$E[\cdot]$	statistical averaging
f	probability density function
F	probability distribution function
$\mathbf{g} = (0,0,-g)^t$	acceleration vector due to gravity
H	wave height
H_s	significant wave height
$\mathbf{K} = (K_X, K_Y)$	wave vector
K	wavenumber
K_p	mean wavenumber
m_n	n^{th} spectral moment
m_n^c	n^{th} central spectral moment
$\mathbf{n}$	unit vector normal to the surface
N	wave action in the kinetic equation
P	pressure
P_a	atmosphere pressure
R	autocorrelation function
S	non-symmetric spectrum
$\hat{S}$	wave spectrum
T	time
T_p	mean wave period
$\mathbf{U} = (U,V,W)^t$	the velocity field
$\mathbf{U}_c$	current velocity
U_n	normal to the surface velocity
U_w	wind velocity
$\mathbf{X} = (X,Y)^t$	horizontal plane coordinate
(X,Y,Z)	coordinates
X_f	fetch
δ_Ω	spectral width
ε	linearization parameter
$\phi(X,Y,Z,T)$	velocity potential
γ	peakedness in the JONSWAP spectrum
γ	skewness
$\eta(X,Y,T)$	surface elevation
κ	kurtosis
λ	wavelength
μ	dynamic viscosity
$\mu \equiv \mu_1$	first statistical moment, the expected value
μ_n	n^{th} statistical moment

μ_n^c	n^{th} central statistical moment
ν	kinematic viscosity
ρ	water density
σ	standard deviation, σ^2 is the variance
ω	vorticity
Ω	cyclic wave frequency
Ω_i	intrinsic frequency
Ω_p	mean wave frequency
∇	gradient operator

References

Baxevani A, Rychlik I (2006) Maxima for Gaussian seas. Ocean Eng 33:895–911
Bitner-Gregersen EM, Magnusson AK (2005) Extreme events in field data and in a second order wave model. In: Olagnon M, Prevosto M (eds) Rogue Waves 2004, Ifremer, France
Cavaleri L (2005) Wave modelling: The present and the future. In: Proc. 14th Aha Huliko'a Winter Workshop, Honolulu, Hawaii, 2005. http://www.soest.hawaii.edu/PubServices/2005pdfs/Cavaleri.pdf. Accessed 15 March 2008
Creamer DB, Henyey F, Schult R, Wright J (1989) Improved linear representation of ocean surface waves. J Fluid Mech 205:135–161
Davidan IN, Lopatukhin LI, Rozhkov VA (1985) Wind waves in the World Ocean. Gidrometeoizdat, Leningrad. (In Russian)
Donelan MA, Magnusson AK (2005) The role of meteorological focusing in generating rogue wave conditions. In: Proc. 14th Aha Huliko'a Winter Workshop, Honolulu, Hawaii, 2005
Draper L (1964) 'Feak' ocean waves. Oceanus 10:13–15
Dysthe K, Krogstad HE, Müller P (2008) Oceanic rogue waves. Annu Rev Fluid Mech 40:287–310
Fenton JD (1979) A high-order cnoidal wave theory. J Fluid Mech 94:129–161
Forristall GZ (2005) Understanding rogue waves: Are new physics really necessary? In: Proc. 14th Aha Huliko'a Winter Workshop, Honolulu, Hawaii, 2005
Gelci R, Cazalé H, Vassal J (1956) Utilisation des diagrammes de propagation à la prevision énergétique de la houle. Bull Inform Comité Central Océanogr Etude Côtes 8:170–187. (In French)
Gelci R, Cazalé H, Vassal J (1957) Prévision de la houle. La méthode des densités spectroangulaire. Bull Inform Comité Central Océanogr Etude Côtes 9:416–435. (In French)
Hasselmann K (1962) On the nonlinear energy transfer in a gravity wave spectrum. Part 1. General theory. J Fluid Mech 12:481–500
Hasselmann K (1968) Weak-interaction theory of ocean waves. In: Holt M (ed) Basic Developments in Fluid Dynamics, vol 2. Academic Press, New York, pp 117–182
Haver S (2005) Freak waves: a suggested definition and possible consequences for marine structures. In: Olagnon M, Prevosto M (eds) Rogue Waves 2004, Ifremer, France
Haver S, Andersen OJ (2000) Freak waves – rare realizations of a typical extreme wave population or typical realizations of a rare extreme wave population? In: Proc. 10th Int Offshore and Polar Eng Conf ISOPE, Seattle, USA, 2000, pp 123–130
Huang NE, Long SR, Tung CC et al (1983) A non-Gaussian statistical model for surface elevation of nonlinear random wave fields. J Geophys Res 88:7597–7606
Janssen PAEM (2003) Nonlinear four-wave interactions and freak waves. J Phys Oceanogr 33:863–884
Janssen P (2004) The interaction of ocean waves and wind. Cambridge University Press, Cambridge

References

Johnson RS (1997) A modern introduction to the mathematical theory of water waves. Cambridge University Press, Cambridge

Kharif C, Pelinovsky E (2003) Physical mechanisms of the rogue wave phenomenon. Eur J Mech B/Fluids 22:603–634

Krasitskii VP (1994) On reduced equations in the Hamiltonian theory of weakly nonlinear surface waves. J Fluid Mech 272:1–30

Lavrenov IV (2003) Wind waves in ocean: dynamics and numerical simulations. Springer-Verlag, Heidelberg

Massel SR (1996) Ocean surface waves: their physics and prediction. World Scientific Publishing Co Pte Ltd, Singapore

Mei CC (1983) The applied dynamics of ocean surface waves. Wiley-Interscience, New York

Mori N, Yasuda T (2002) A weakly non-Gaussian model of wave height distribution for random wavetrain. Ocean Eng 29:1219–1231

Müller P, Osborne A, Garrett C (2005) Rogue waves. Oceanography 18:66–75

Olagnon M, Magnusson AK (2004) Sensitivity study of sea state parameters in correlation to extreme wave occurrences. In: Proc. 14th Int Offshore and Polar Eng Conf ISOPE, Toulon, France, 2004, pp 18–25

Petrova P, Cherneva Z, Guedes Soares C (2007) On the adequacy of second-order models to predict abnormal waves. Ocean Eng 34:956–961

Pierson WJ, Moskowitz L (1964) A proposed spectral form for fully developed wind seas based on the similarity theory of S.A. Kitagorodskii. J. Geophys. Res. 69:S181–S190

Polnikov VG (2007) Nonlinear theory of random wave fields on water. Lenand, Moscow. (In Russian)

Prevosto M (2001) Statistics of wave crests from second order irregular wave 3D models. In: Olagnon M, Athanassoulis GA (eds) Rogue Waves 2000, Ifremer, France, pp 59–72

Prevosto M, Bouffandeau B (2002) Probability of occurrence of a "giant" wave crest. In Proc 21st Int Conf OMAE 2006, Oslo, Norway, 2002, OMAE2002-28446:1-8

Rosenthal W (2005) Results of the MAXWAVE project. In: Proc. 14th Aha Huliko'a Winter Workshop, Honolulu, Hawaii, 2005. http://www.soest.hawaii.edu/PubServices/2005pdfs/Rosenthal.pdf. Accessed 14 March 2008

Socquet-Juglard H, Dysthe KB, Trulsen K, Krogstad HE, Liu J (2005) Probability distributions of surface gravity waves during spectral changes. J Fluid Mech 542:195–216

Stansell P (2004) Distributions of freak wave heights measured in the North Sea. Appl Ocean Res 26:35–48

Tayfun M (1980) Narrow–band nonlinear sea waves. J Geophys Res 85 C3:1548–1552

Tung CC, Huang NE (1985) Peak and trough distributions of nonlinear waves. Ocean Eng 12:201–209

Whitham GB (1974) Linear and nonlinear waves. Wiley & Sons, New York London Sydney Toronto

Zakharov V (1968) Stability of periodic waves of finite amplitude on a surface of deep fluid. J Appl Mech Tech Phys 2:190–194

Zakharov V (1999) Statistical theory of gravity and capillary waves on the surface of a finite-depth fluid. Eur J Mech/B – Fluid 18:327–344

Zakharov VE (1974) The Hamiltonian formalism for waves in nonlinear media having dispersion. Radiofizika 17:431–453. (In Russian) [(1975) Radiophys Quantum Electronics 17:326–343]

Zakharov VE, Kuznetsov EA (1997) Hamiltonian formalism for nonlinear waves. Physics-Uspekhi 40:1087–1116

Zhang J, Yang J, Wen J, Prislin I, Hong K (1999) Deterministic wave model for short-crested ocean waves: Part I. Theory and numerical scheme. Appl Ocean Res 21:167–188

Chapter 3
Quasi-Linear Wave Focusing

The real sea surface may be described as the superposition of many wave packets propagating in different directions with various speeds. Collisions of such packets can lead to significant time and space variability of the wave field, including the amplification of the wave energy in some areas. Various mechanisms can provide growth of the wave amplitude: geometrical convergence of the wave fronts in shallow water and above underwater sills, intersection of wave packets propagating with different speeds and directions, wave refraction on oceanic currents, etc. Variable wind and atmospheric pressure above the sea induce nonuniform distribution of the water-wave field contributing to the process of the freak-wave formation. A description of these mechanisms is given in the following sections within the framework of the linear theory, demonstrating the main features of the generation of rogue waves, including their short-lived character and random occurrence.

3.1 Geometrical Focusing of Water Waves

Amplification of water waves due to the effect of geometrical focusing is a well-known process for waves of any physical nature. Geometrical focusing, as it may be concluded from the title, is related to spatial variability of wave fronts and/or medium parameters. Directional sea-wave distribution occurs when the waves come from different directions: for the open sea, from several stormed areas, and for the coastal zone, due to the reflection from complicated coastal lines. Alternatively, wave fronts become curved when propagating over basins with a variable seafloor. Figure 3.1 illustrates how waves approaching from the open sea undergo deformation of wave fronts due to bottom topography near the coast; in the end, the fronts come into alignment with the shoreline. The refracted waves may interfere, providing wave energy concentration near capes.

A simplified model of geometrical focusing of linear water waves is based on the ray theory, assuming that the wavelength is smaller than the characteristic length scales of the bottom variability and curvature radius of the wave fronts. In this case, the monochromatic progressive plane wave can be described locally as

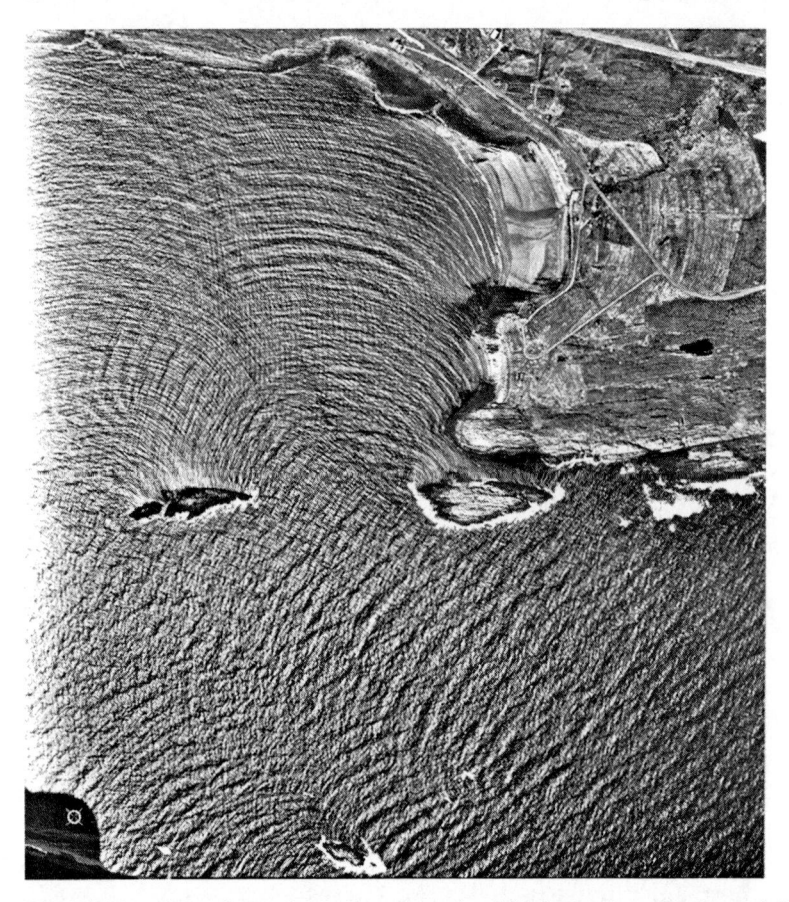

Fig. 3.1 Wave fronts change due to obstacles and variable bathymetry (Dysthe et al. 2005)

$$\eta(\mathbf{X},T) = A(\mathbf{X},T)\exp[i\theta(\mathbf{X},T)] + c.c., \qquad (3.1)$$

where *c.c.* denotes complex conjugation, and the wave frequency Ω and wave vector $\mathbf{K}$ are determined through the phase θ

$$\Omega(\mathbf{X},T) = \frac{\partial \theta}{\partial T}, \quad \mathbf{K}(\mathbf{X},T) = -\nabla\theta. \qquad (3.2)$$

Here, the vector $\mathbf{X} = (X, Y)$ is in the horizontal plane (the plane sea surface, see Fig. 2.1), and $\mathbf{K} = (K_X, K_Y)$ is the wave vector. All the wave parameters (amplitude, wave frequency, and wave vector) are assumed to be slowly varying functions of time and space (when compared with characteristic wave period and wave length). From the definition of the wave frequency and wavenumber (3.2), the following kinematic equations can be derived (Whitham 1974, Ostrovsky and Potapov 1999).

$$\frac{\partial \mathbf{K}}{\partial T} + \nabla\Omega = 0, \quad \nabla \times \mathbf{K} = 0. \qquad (3.3)$$

3.1 Geometrical Focusing of Water Waves

System (3.3) is closed and gives the irrotational field of the wave vector **K** when adding the dispersion relation (obtained in Chap. 2) for water waves in basins with slowly varying bathymetry

$$\Omega = \sqrt{gK\tan(KD)}, \tag{3.4}$$

where K is the modulus of **K**. Here, we consider an ocean with still water; the case of oceanic current will be analyzed further in Sect. 3.4.

The wave amplitude can be found from the equation of energy balance, which can be derived from the Euler equation, assuming smooth variation of wave parameters (Whitham 1974, Ostrovsky and Potapov 1999)

$$\frac{\partial A^2}{\partial T} + \nabla \cdot (C_{gr} A^2) = 0, \tag{3.5}$$

where the group velocity $C_{gr} = \partial \Omega / \partial K$ is determined from the dispersion relation (3.4). This system of Eqs. (3.3, 3.4, 3.5) allows us to find all characteristics of the wave field if the initial wave parameters are known. In general, they describe the spatio-temporal evolution of the wave field and will be used in this chapter in applications of various particular cases of wave evolution.

To emphasize the main effects of geometrical focusing, let us consider waves of constant frequency in a basin of variable depth. In this case, the kinematic equations can be given in the following Hamiltonian form (Brekhovskikh 1980, Ostrovsky and Potapov 1999):

$$\frac{d\mathbf{X}}{dT} = \frac{\partial \Omega}{\partial \mathbf{K}}, \quad \frac{d\mathbf{K}}{dT} = -\frac{\partial \Omega}{\partial \mathbf{X}}. \tag{3.6}$$

These equations determine the ray pattern on the sea surface. The equation of energy balance (3.5) is reduced for monochromatic waves to the energy flux conservation along the rays (Brekhovskikh 1980, Ostrovsky and Potapov 1999)

$$C_{gr} b A^2 = \text{const}, \tag{3.7}$$

where $b(X,Y)$ is the distance between neighboring rays (the differential ray width) found after solving ray equation (3.6).

Trivial ray patterns formed in basins of infinite depth (deep water), when the dispersion relation is $\Omega = (gK)^{1/2}$, do not include depth variation. In this case, the whole rays are straight lines

$$Y - Y_0 = \gamma(X - X_0), \tag{3.8}$$

where X_0 and Y_0 are initial coordinates of the ray and γ is the slope of the ray. In general, the rays are not parallel lines, and the distance between them can vary, influencing the wave amplitude. It is evident that the maximum amplification can be achieved if the whole rays converge in one point (focus)—for instance, the cylindrical focusing wave whose amplitude is proportional to $R^{-1/2}$ according to (3.7)—since the distance between rays is proportional to the radius, R. Such a focusing is very often described in textbooks on optics.

Fig. 3.2 Location of focal points for various curvatures of focusing cylindrical waves

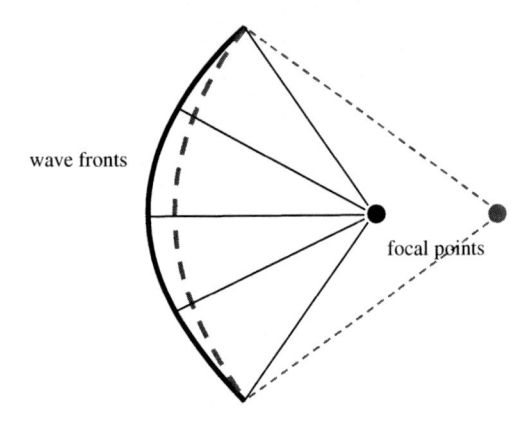

A purely cylindrical focusing wave seems to be an unrealistic model of wind waves in the open sea, and the geometrical focusing of waves can be only due to a complicated structure of storm areas, as it is shown in Fig. 3.2. Meanwhile, this very simplified model can explain rogue wave formation due to geometrical focusing. Changing conditions of wave generation influences the curvature of the wave front, and the focus point shifts as a result (see demonstration in Fig. 3.2), or may even split into spatially distributed focusing areas called caustics, if the wave front is not cylindrical (Fig. 3.3).

Although in the framework of ray theory the wave amplitude goes to infinity when two neighboring rays intersect, more accurate theories show that the amplitude is finite in all cases except for a purely cylindrical wave (Brekhovskikh 1980, Arnold 1990). The wave amplitude in a focus area is fairly small compared to that in a focus point and can be very nonuniform and variable along the caustics. The wind variability in storm areas should provide irregular formation of focuses and caustics at different places. Hence, this mechanism can explain the unpredictable character of short-lived, large-amplitude wave (rogue wave) occurrence in constant depth water due to spatio-temporal variability of the wind flow and atmospheric pressure above waves. This scenario of rogue-wave generation in the open ocean requires the existence of focusing wave fronts in storm areas.

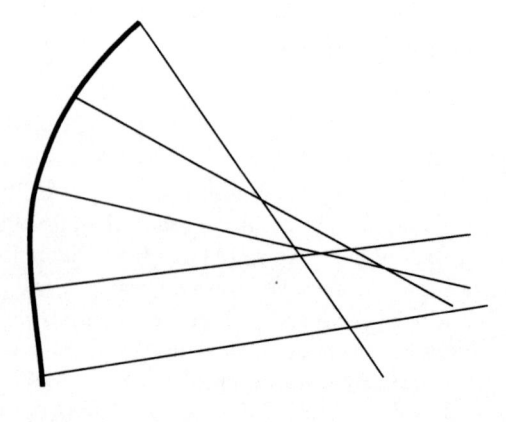

Fig. 3.3 Spatially distributed focal area when the initial wave front is not cylindrical

3.1 Geometrical Focusing of Water Waves

Another situation is realized for waves in shallow water where the bottom variability plays a significant role. The rays are no longer straight lines, and a lot of caustic curves and focal points can be formed. Below, we give an analytical example of ray patterns in shallow water with parallel isobaths, with depth being a function of one coordinate, $D = D(X)$. In this case, the alongshore component of the wave vector is constant (K_Y = const). The wavenumber is found in explicit form from the shallow-water approximation of the dispersion relation (3.4) as $K = \Omega/(gD)^{1/2}$. The well-known Snell law for the angle between ray and isobath (the slip angle θ)

$$\frac{\sin\theta(X)}{\sin\theta_0} = \sqrt{\frac{D(X)}{D_0}} \tag{3.9}$$

(index "0" corresponds hereafter to the initial values) follows from ray equation (3.6). As a result, the ray trajectory is given by

$$Y(X) - Y_0 = \int_{X_0}^{X} \frac{d\xi}{\sqrt{[D_0/D(\xi)] - \cos^2\theta_0}}. \tag{3.10}$$

In the particular case of a parabolic bottom $D(X) = D_0 \cdot (X/X_0)^2$, the rays are arcs of a circle

$$(Y - Y_0 - X_0 \tan\theta_0)^2 + X^2 = \left(\frac{X_0}{\cos\theta_0}\right)^2. \tag{3.11}$$

The center is located on the shoreline, and the radius depends on the slip angle. From this simple analytical example, some general properties of rays in basins of variable depth can be obtained: (i) rays turn towards the shore; (ii) wave reflection from deep water; (iii) forming of waveguides above underwater ridges; (iv) edge wave existence on bottom slopes; (v) caustic and focus occurrence. All of them are often observed in nature and are present partly in Fig. 3.1.

The ray dynamics in basins of variable depth, even if the bottom topography is regular and simple, can be very complicated (see Dobrokhotov et al. 2006), including random behavior. In this case, Hamiltonian methods can be very effective (Abdullaev and Zaslavsky 1993). As a result, the wave field in real shallow sea contains caustics where the wave field is intensified. Figure 3.4 shows the computed ray patterns of long waves in the Sea of Japan induced by an isotropic source of a circlular shape (taken from Choi et al. 2002), demonstrating the random distribution of focal points in the sea. Formally, caustics here are time independent, although in reality location of the caustics and focuses becomes random due to wind flow variability in storm areas. The focal areas in some certain places may appear and disappear quickly, emphasizing the randomness and short-lived character of rogue waves.

As pointed out above, the wave amplitude in the framework of ray theory tends to infinity at caustics, and the adequate theory should be beyond the approximation of slowly varying wave amplitude. General mathematical ray theory in homogeneous and inhomogeneous media, including the classification of caustics, is

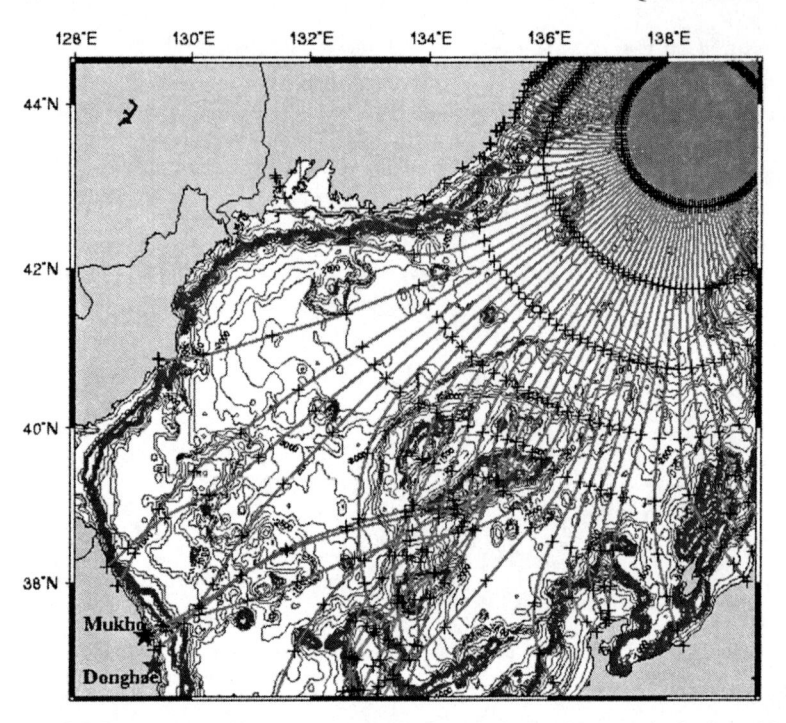

Fig. 3.4 Ray patterns of long waves traveling from an isotropic source in the Sea of Japan

well developed (see Arnold 1990, Babic and Buldyrev 1991, Brekhovskikh 1980, Dobrokhotov and Zhevandrov 2003, Mei 1983). The wave field in the vicinity of caustics is bounded—in particular, for simplified caustics the amplification factor is proportional to $(KL)^{1/6}$, where L is the characteristic scale of the bottom variation. Its value is not too large, so in the context of rogue waves, only few caustics can satisfy the amplitude criterion of rogue waves (amplification by a factor two and more (I.1)). As a result, the statistics of focuses and caustics that can be obtained from the ray equation (3.6) overestimate the probability of freak wave occurrence and has to be calculated in the framework of more accurate theories. Until now, this problem has not been solved.

Thus, the linear theory of the monochromatic water-wave propagation in a basin of variable depth demonstrates that the bottom variability in shallow seas and coastal zones, together with the variability of the air flow in storm areas, can be an effective "generator" of the rogue wave phenomenon. Meanwhile, these effects cannot accurately specify the precise position and time of the rogue-wave occurrence, although some forecasting on the basis of the weather conditions is potentially possible.

Nonlinearity certainly modifies the process of the geometrical focusing of water waves. Firstly, nonlinearity generates high harmonics, which satisfies the dispersion relation, and thus freely propagates with velocities different from that of the basic wave (free harmonics). This part of wave energy is scattered in space modifying the amplitude of the freak wave. Secondly, nonlinearity can lead to wave breaking, which modifies the wave amplitude at caustics. And thirdly, nonlinearity changes the speed of propagating waves modifying the location of focal points and wave amplitude at these points. Nonlinear effects on caustics have been analyzed theoretically for deep water in Peregrine and Smith (1979) and Peregrine (1983) and for shallow water in Engelbrecht et al. (1988) and Pelinovsky (1982). They showed that nonlinearity actually cannot completely destroy the process of geometrical focusing of the wave field, and therefore large-amplitude rogue waves can appear in the ocean due to this mechanism. Nonlinear study of the freak wave formation will be studied in Chaps. 4 and 5 in detail.

3.2 Dispersive Enhancement of Wave Trains

The spatial heterogenity of the wave field, which results from geometrical focusing, is accompanied by nonuniformity of the wave trains represented by the frequency or spatial spectrum (see Chap. 2). Due to strong dispersion of water waves, each individual sine wave travels with a frequency-dependent velocity, and may travel along different directions. The interference pattern of many sine waves with different frequencies can become intricate. At one moment, short waves with small group velocities are located in front of long waves with large group velocities, but then after some time the long waves will overtake the shorter waves and a large-amplitude wave can occur due to the spatio-temporal superposition. Draper (1964) perhaps first suggested this idea as a possible rogue wave generation mechanism. Afterwards, the long waves turn out to be in front of the short waves, and the amplitude of the wave train decreases due to the spreading of the wave train. It is obvious that a significant focusing of the wave energy can occur only if the waves with different lengthscales merge at a fixed location at the same time. Specific locations of transient wave groups with different length (and time) scales in storm areas can appear. Let us consider the case of a freshening wind as an example. Due to the resonant character of the wave generation by wind, a light wind generates short waves while an increasing wind generates longer waves. In this way, the dispersive nature of water waves may cause rogue-wave formation even in the case of unidirectional waves when they are frequency modulated.

Unidirectional transient wave groups will be analyzed in detail hereafter. To emphasize the dispersive focusing of unidirectional water waves, the kinematic equation (3.3) derived in Sect. 3.1 can be reduced to a single equation describing the spatio-temporal evolution of the characteristic wave frequency, Ω (Whitham 1974, Ostrovsky and Potapov 1999):

$$\frac{\partial \Omega}{\partial T} + C_{gr}(\Omega)\frac{\partial \Omega}{\partial X} = 0. \tag{3.12}$$

For the sake of simplicity, we assume constant water depth. Multiplying by $\partial C_{gr}/\partial \Omega$ Eq. (3.12) transforms into the universal form

$$\frac{\partial C_{gr}}{\partial T} + C_{gr}\frac{\partial C_{gr}}{\partial X} = 0. \tag{3.13}$$

This partial differential equation of the first order is equivalent to the system of ordinary differential equations

$$\frac{dX}{dT} = C_{gr}(\Omega), \quad \frac{dC_{gr}}{dT} = 0 \tag{3.13a}$$

describing an evident physical feature, and each spectral wave component propagates with its own group velocity. The solution of the quasi-linear hyperbolic equation (3.13) or the equivalent system (3.13a) corresponds to a simple (Riemann) wave

$$C_{gr}(X,T) = C_0(\xi) = C_0(X - C_{gr}T) \tag{3.14}$$

where $C_0(\xi)$ corresponds to the initial spatial distribution of the wave groups with different frequencies (and group velocities). The form of such a kinematic wave transforms continuously with distance (time), as is shown in Fig. 3.5. The slope of the group velocity curve can be calculated directly from (3.14)

$$\frac{\partial C_{gr}}{\partial X} = \frac{dC_0/d\xi}{1 + TdC_0/d\xi}. \tag{3.15}$$

If long waves are located in front of short waves, $dC_0/d\xi > 0$, and the slope $\partial C_{gr}/\partial X$ decreases with time, what reflects the increase of distance between long and short waves. The case $dC_0/d\xi < 0$ (or $dC_0/dX < 0$ at $T = 0$) corresponds to long waves being placed behind short waves. The process of long waves overtaking short waves corresponds to the initial increase of the slope of the kinematic wave up to infinity, followed by a decrease. The first merging of several wave groups with neighboring slightly different frequencies at the same point (wave focusing) occurs at time

$$T_f = \frac{1}{\max[-dC_0/dX]}. \tag{3.16}$$

It is obvious that several focusing points are possible for a transient wave group depending on the frequency distribution. The case corresponding to all wave groups merging at the same point, X_f, and time, T_f, is described by the self-similar solution of Eq. (3.13)

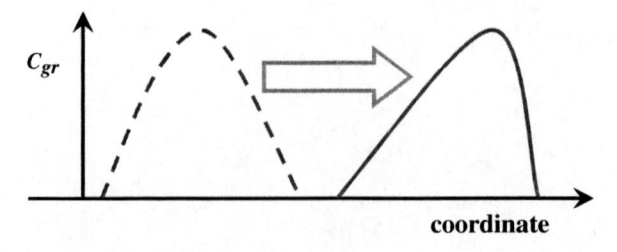

Fig. 3.5 A qualitative picture of a group-velocity wave curve deformation due to disperion

Fig. 3.6 Deformstion of a group-velocity curve in case of a wave train with the linear frequency modulation

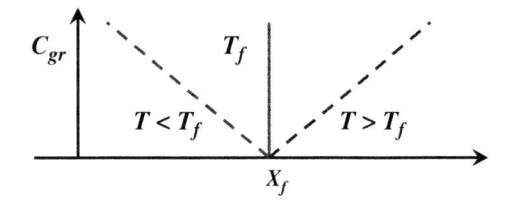

$$C_{gr} = \frac{X - X_f}{T - T_f}, \qquad (3.17)$$

which is illustrated in Fig. 3.6.

The interval of all possible water wave group velocities is from $C_{LW} = (gD)^{1/2}$ (the fastest, long waves) up to zero (if capillary effects are neglected) (see Fig. 2.3). According to (3.17) wave energy from spatial domain $C_{LW}T_f$ compresses into a null area when $T = T_f$ (the concentration point). The required variation of the wave frequency (or constrained wave number) in the group for maximum (optimal) focusing can be easily found from Eq. (3.17). In the deep-water limit $C_{gr} = g/(2\Omega)$, performing such an optimal focusing of a paddle in the deep-water laboratory tank should generate a wave train with the following frequency variation (it can be easily obtained from Eq. (3.17) for $X = 0$):

$$\Omega_{opt}(T) = \frac{g(T_f - T)}{2X_f}. \qquad (3.18)$$

The wave amplitude satisfies the energy balance equation (3.5), of which the solution in the one-dimensional case is found explicitly to be

$$A(X,T) = \frac{A_0(\xi)}{\sqrt{1 + T(dC_0/d\xi)}}, \qquad (3.19)$$

where $A_0(\xi)$ is the initial spatial distribution of wave amplitude. At each focal point, the wave becomes extreme, having infinite amplitude; near the focal point, the amplitude obeys the asymptotic law $A \sim (T_f - T)^{-1/2}$.

Noticing that each realization of wind waves always turns into frequency- and amplitude-modulated wave groups, and that the kinematic approach predicts infinite wave height at caustic points, the probability of freak-wave occurrence should be very high. In fact, the situation is more complicated and less dramatic when more accurate theories are employed. The kinematic approach assumes slow variations of the amplitude and frequency (group velocity) along the wave group, although this assumption breaks down in the vicinity of focal points (possible limitations of the wave amplitude due to nonlinear effects and wave breaking will not be discussed in this section). It is a well-known problem in ray theory as a whole, not only regarding water waves. Generalizations of the kinematic approach in the linear theory can be done by using various expressions of the Fourier integral for the wave field near the caustics. In a generalized form, it has been expressed through the Maslov integral representation, described in detail for water waves by Dobrokhotov (1983),

Lavrenov (2003), Brown (2000, 2001) and Dobrokhotov and Zhevandrov (2003). In particular, Brown (2001) pointed out the relationship between the focusing of a unidirectional wave field and "canonical" caustics: fold and longitudinal cusp. We consider here the simplified form of such a representation for the conditions of optimal focusing (3.17), and use the standard form of the direct and inverse Fourier transformation for the water wave displacement in a hydrodynamic flume,

$$\eta(X,T) = \int_{-\infty}^{+\infty} \eta(\Omega)\exp\left[i(\Omega T - KX)\right]d\Omega, \tag{3.20}$$

$$\eta(\Omega) = \frac{1}{2\pi}\int_{-\infty}^{+\infty} \eta_0(T)\exp(-i\Omega T)dT, \tag{3.21}$$

where $\eta_0(T) = \eta(X=0,T)$ is the water displacement generated by the paddle, and the wave frequency and wavenumber satisfy the dispersion relation (3.4). Hence, the spatial wave evolution will be considered hereafter for a given boundary condition $\eta(X=0,T)$. This way is convenient for laboratory experiments.

3.2.1 Exact Solution for the Delta-Function

First of all, the physical problem of rogue wave formation may be reformulated in mathematical terms as the problem of singularity occurrence from smooth initial data. Due to invariance of the Fourier integral with respect to changes of signs of the coordinate, X, and time, T, this problem has a straightforward link to the mathematical theorem of smooth solutions of the Cauchy problem for singular initial data. This is the case for water waves; a singular Dirac delta-function disturbance (a rogue wave prototype) transforms into a smooth wave field (Green's function). The corresponding analytical expressions can be found in the limiting cases of deep and shallow water. In the general case, the wave field far from the paddle can be described by the asymptotic solution using the method of stationary phase

$$\eta(X,T) = Q\sqrt{\frac{1}{2\pi X \cdot |d^2K/d\Omega^2|}}\cos\left(\Omega T - KX - \pi/4\right), \tag{3.22}$$

where the group velocity, C_{gr} (and also the wave frequency and the wave number) is defined by the condition of optimal focusing (3.17) for a fixed coordinate, X (far from the paddle it has a simple form: $C_{gr} = X/T$). The parameter Q in (3.22) defines the intensity of the delta function.

When the deep-water condition is satisfied, expression (3.22) reduces to

$$\eta(X,T) = Q\sqrt{\frac{g}{2\pi X}}\cos\left[\frac{gT^2}{4X} - \frac{\pi}{4}\right] \tag{3.23}$$

and describes a frequency-modulated wave train with an amplitude decreasing with distance as $X^{-1/2}$. In the vicinity of the leading wave ($K \to 0$), expressions (3.23) and (3.22) are not valid (the wavelength becomes comparable to the distance to the source) and should be replaced by

$$\eta(X,T) = Q \left(\frac{2}{XD^2}\right)^{1/3} \mathrm{Ai}\left[\left(\frac{2}{XD^2}\right)^{1/3}(C_{LW}T - X)\right], \quad (3.24)$$

derived from (3.20) by using the long-wave approximation of the dispersion relation,

$$\Omega = C_{LW}\left(1 - \frac{K^2 D^2}{6}\right), \quad C_{LW} = \sqrt{gD}. \quad (3.25)$$

Here, $\mathrm{Ai}(\xi)$ is the Airy-function. As a result, the amplitude of the leading wave decreases as $T^{-1/3}$, and its period (duration) increases as $X^{1/3}$.

Thus, the delta-function disturbance transforms into a smooth wave field, and owing to invariance with respect to coordinate and time, one may say that an initially smooth wave field in the forms (3.23) and (3.24), with inverted coordinate and time, evolves into a freak wave of infinite height. These solutions demonstrate what kind of wave packets can generate a freak wave of delta-like shape in the process of wave evolution. Bona and Saut (1993) have shown that a singularity (dispersive blowup) can be achieved in the long-wave approximation from the following initial continuous function, having a finite energy integral ($1/8 < m < 1/4$)

$$\eta(X=0,T) \propto \frac{\mathrm{Ai}(t)}{(1+t^2)^m}, \quad (3.26)$$

where t is a scaled time (see Bona and Saut 1993).

3.2.2 Exact Solution for a Gaussian Wave Train

Strictly speaking, singular solutions of linearized equations are limited by mathematical applications only. Meanwhile, the integral (3.20) has been calculated for other smooth shapes suggested as potential freak waves. An exact analytical solution exists for a Gaussian envelope over deep water (Clauss and Bergmann 1986, Clauss 1999, Magnusson et al. 1999). It has the form

$$\eta(X,T) = \frac{A_0}{(1+16\Omega_{env}^4 X^2/g^2)^{1/4}} \exp\left(-\frac{\Omega_{env}^2}{1+16\Omega_{env}^4 X^2/g^2}(T-X/C_{gr})^2\right)$$

$$\times \cos\left(\frac{\Omega_0(T-X/C_{ph})}{1+16\Omega_{env}^4 X^2/g^2} + \frac{4\Omega_{env}^4 XT^2}{g(1+16\Omega_{env}^4 X^2/g^2)} - \frac{1}{2}\mathrm{atan}\left[\frac{4\Omega_{env}^2 X}{g}\right]\right), \quad (3.27)$$

where A_0 is the wave train amplitude, and Ω_{env} and Ω_0 are frequencies of the wave envelope and carrier wave, respectively, at the location of the flume ($X = 0$). The

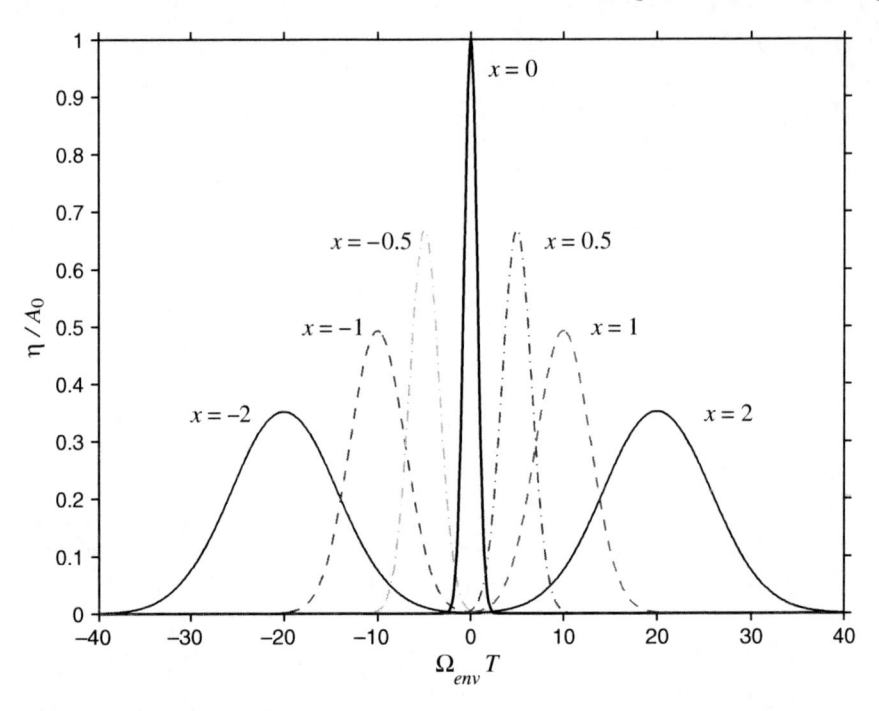

Fig. 3.7 Evolution of the envelope of a Gaussian transient group (3.27). The distance is given in normalized variables $x = X\Omega_{env}^2/g$. The carrier wave frequency is chosen as $\Omega_0 = 5\Omega_{env}$

initial pulse has the shape of a delta function when $\Omega_{env} \to \infty$, and then solution (3.27) transforms into (3.23). When coordinate and time are inverted (more precisely, replacing X by $(X_f - X)$ and T by $(T_f - T)$), the solution (3.27) describes a wave train focusing with increasing amplitude (for $X < X_f$), and then defocusing with decreasing amplitude (for $X > X_f$). The evolution of the envelope amplitude is shown in Fig. 3.7. Both processes are readily observed: wave focusing and defocusing.

A similar solution of the Cauchy problem may be found in the long-wave approximation for a Gaussian pulse-like initial perturbation with amplitude A_0 and duration Ω^{-1} (Pelinovsky et al. 2001)

$$\eta(X = 0, T) = A_0 \exp(-\Omega^2 T^2). \tag{3.28}$$

The corresponding solution for $X > 0$ is given by

$$\eta(X,T) = \frac{A_0 C_{LW}}{\Omega \sqrt[3]{\frac{D^2 X}{2}}} \exp\left\{\frac{C_{LW}^2}{2D^2 X \Omega^2}\left(C_{LW}T - X - \frac{6C_{LW}^4}{77D^2 X \Omega^4}\right)\right\}$$

$$\times \mathrm{Ai}\left\{\left(C_{LW}T - X - \frac{9C_{LW}^4}{77D^2 X \Omega^4}\right)\left(\frac{D^2 X}{2}\right)^{-\frac{1}{3}}\right\}. \tag{3.29}$$

3.2 Dispersive Enhancement of Wave Trains

Fig. 3.8 Formation of a rogue wave of Gaussian shape in shallow water: before ($x = -100$), the focusing moment ($x = 0$) and after it ($x = 100$). Dimensionless variables are defined as $t = \Omega T - X\Omega/C_{LW}$, $x = XD^2\Omega^3/(2C_{LW}^3)$

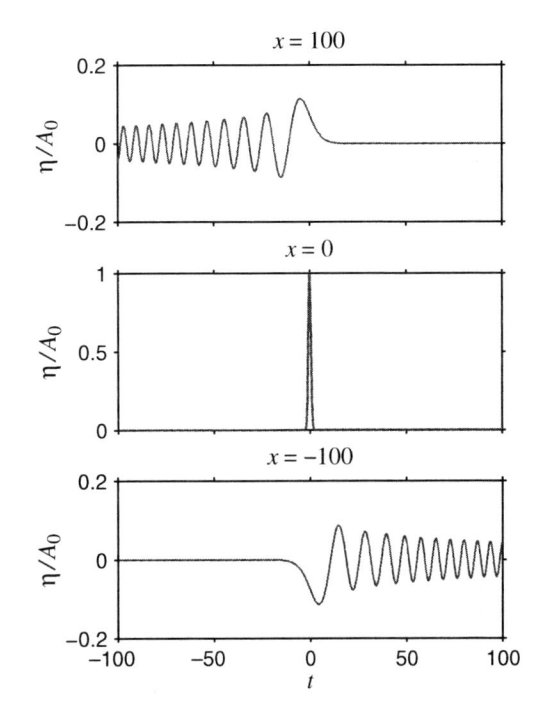

Similar to the case considered above, when inverting the spatial coordinates and time, the wave packet (3.29) forms a Gaussian pulse (3.28) and then again disperses according to (3.29). This evolution that provides a huge wave formation from a dispersive wave packet on shallow water is given in Fig. 3.8.

Exact solutions may be useful for seakeeping tests and freak wave design simulations in ocean engineering. It is evident that in the framework of linear theory it is easy to reproduce a freak wave of any desired shape: asymmetric crest, hole in the sea, wave having a steeper forward face preceded by a deep trough (such shape is often reported in some descriptions of rogue waves—see for instance, Lavrenov 1998, 2003).

It is important to emphasize that the considered mechanism of dispersive focusing is the result of phase coherence of spectral components of wave groups (in-phase superposition). The occurrence of freak waves in random fields represented as the superposition of Fourier components with random phases is more realistic. However, its description is trickier, although easily reproduced within the framework of the linear theory. Nonlinear effects can modify dispersive focusing due to the same reasons as discussed at the end of Sect. 3.1 (Johannessen and Swan 1997, Clauss 1999, Pelinovsky et al. 2000, Kharif et al. 2001, Slunyaev et al. 2002, Goulitski et al. 2004). The effects of nonlinearity and randomness will be considered in Chaps. 4 and 5.

Experiments on the focusing of unidirectional transient groups in flumes have been performed repeatedly during the two last decades (Baldock et al. 1996, Brown and Jensen 2001, Clauss 1999, 2002, Contento et al. 2001, Johannessen and Swan 1997, Stansberg 2001, Goulitski et al. 2004, Touboul et al. 2006, Kharif et al. 2008). In most of them, the transient wave group in deep-water conditions is generated mechanically by a paddle with frequency varying linearly, following (3.18). The process of wave focusing and defocusing is very pronounced, as shown in Fig. 3.9. Effects of nonlinearity and finite depth decrease the amplified wave amplitude predicted by the above considered model solutions. Hence, special correction

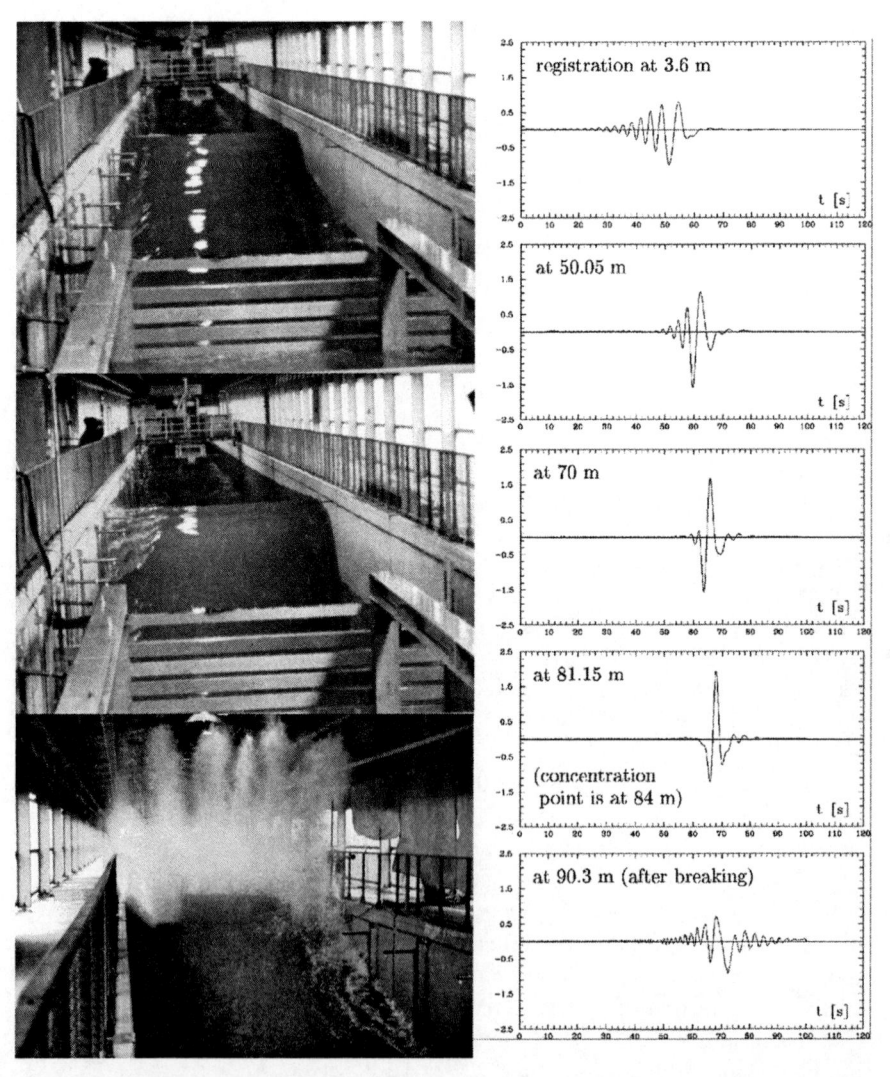

Fig. 3.9 Generation of a high-amplitude (3.2 m) rogue wave from a transient group (water depth 4 m). The graphics show surface elevation in meters. (Clauss 2002, reproduced with permission from Elsevier)

3.2 Dispersive Enhancement of Wave Trains

procedures are applied to provide more optimal conditions for wave focusing in laboratory or numerical wave tanks (see Clauss 2002, Bonnefoy et al. 2005). As a result, the generated huge waves can achieve large amplitudes and may even break (see Fig. 3.9). The amplitude of the rogue wave (3.2 m) is comparable to the water depth (4 m) in this experiment. The main conclusion that eventually follows from the multiple experiments, is that the process of the dispersive focusing, simply explained within the framework of the linear theory, is still valid for nonlinear wave groups, including large-amplitude groups; the effect of dispersive focusing turns out to be a robust mechanism.

In the real 3D ocean, both quasi-linear focusing effects (geometrical and dispersive) may be important and can act supplementarily. The number of 3D experiments is very limited because 3D wave tanks are very costly to operate, and a directionally spread wave generation represents a hard task. Meanwhile, experiments by Johannessen and Swan (2001) have shown that curved wave fronts lead to 3D breaking waves; and the wave amplification can be very large. The same results are achieved in numerical simulations with cylindrical transient wave groups in the framework of 3D fully nonlinear hydrodynamic models (Bateman et al. 2001, 2003; Brandini and Grilli 2001a,b; Fochesato et al. 2007), see Fig. 3.10. The laboratory and numerical experiments confirm that nonlinearity modifies the process of wave focusing, however, without destroying it.

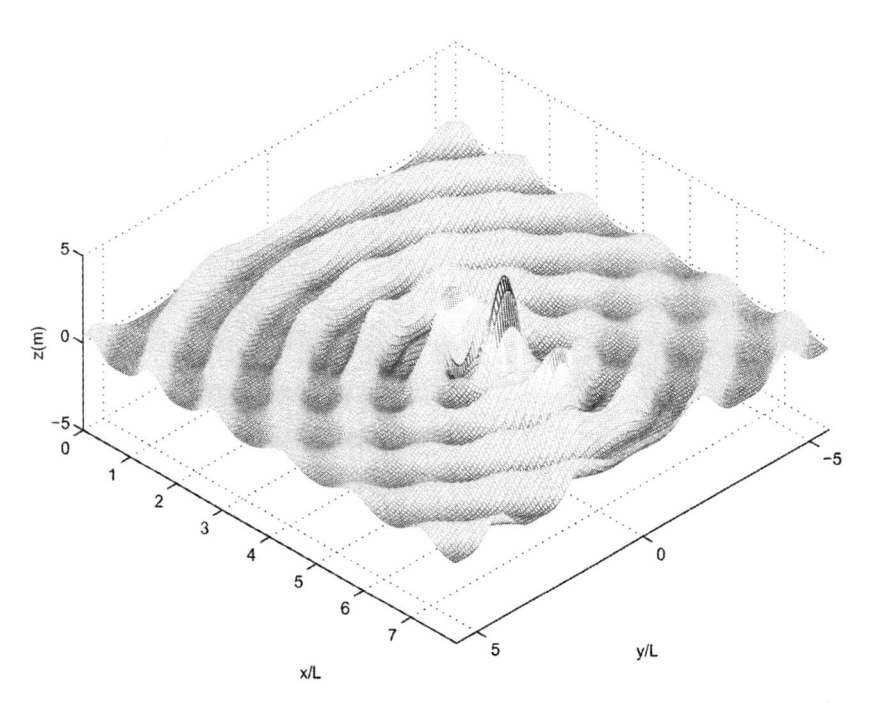

Fig. 3.10 Focusing of a cylindrical transient wave group. Reproduced from Brandini and Grilli (2001b)

3.3 Wave Focusing Under the Action of Wind

Geometrical and dispersive focusing mechanisms considered in the previous sections require spatial and temporal inhomogeneities of the initial distribution of the wave parameters that are provided by inhomogeneous and unsteady wind flow and atmospheric pressure in storm areas. Due to large fetch distances, the wind flow may influence the wave dynamic and kinematic properties. This process was recently studied experimentally in the large wind-wave tank (40 m long, 2.6 m wide, 1 m deep) of IRPHE, Marseille (Touboul et al. 2006, Kharif et al. 2008). For more details see Sect. 4.5.

A paddle can generate regular or random waves in a frequency range of 0.5–2 Hz. For optimal focusing (according to (3.18)), the wave frequency is varied linearly from 1.3 Hz to 0.8 Hz during 10 s; that corresponds to a focusing length of 17 m and a focusing time of 26 s. Figure 3.11 shows a wave focusing at 20 m when no wind is blowing over the water waves. The slight difference existing between the theoretical and experimental values of X_f is due to the nonlinearity of the experimental wave train. Figure 3.12 describes the wave-focusing mechanism for a wind with speed of 6 m/s. The fetch is defined as the distance between the probes on the trolley and the end of the upstream beach where air flow meets the water surface (see Sect. 4.5).

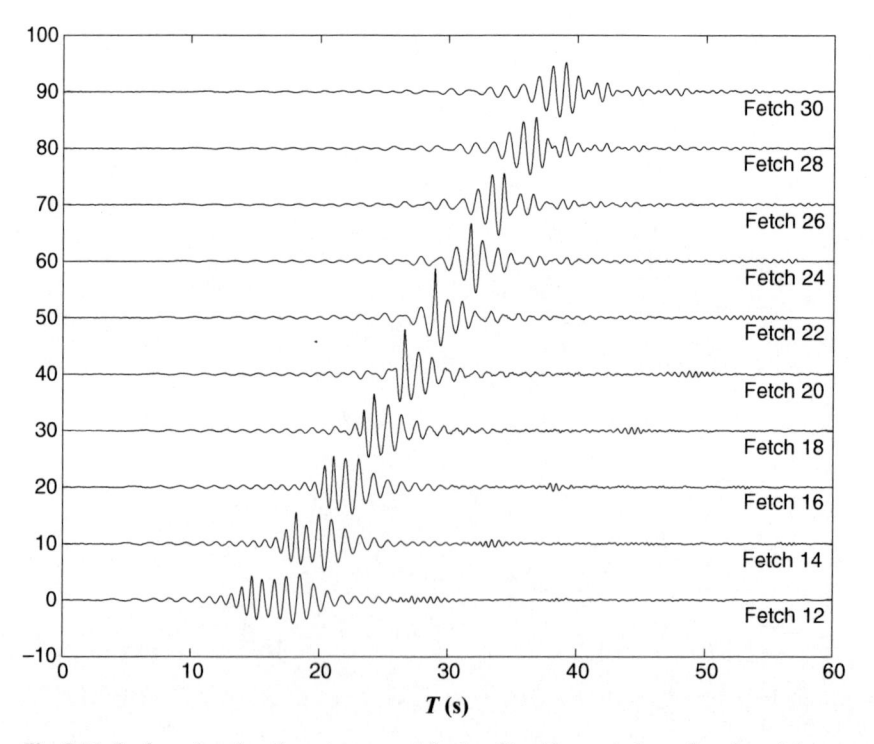

Fig. 3.11 Surface elevation (in cm) at several fetches X, without wind as a function of time

3.3 Wave Focusing Under the Action of Wind

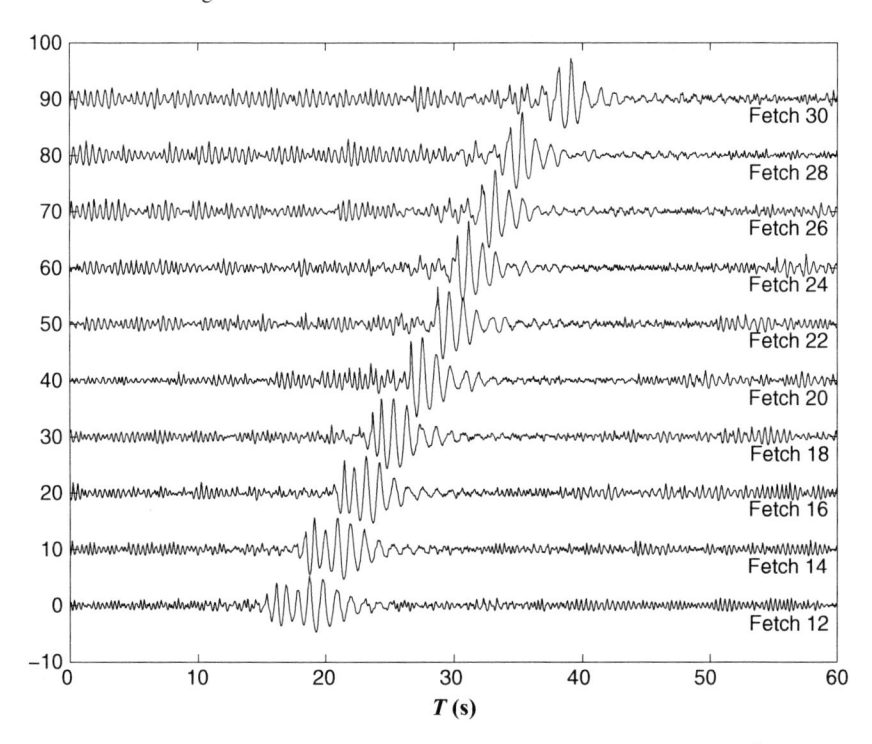

Fig. 3.12 Surface elevation (in cm) at several fetches X, for wind speed $U_w = 6 \, \text{m/s}$, as a function of time

It is clearly seen that wave focusing looks qualitatively similar to the case with no wind, except for the presence of noise due to direct generation of small-scale waves. Discussion of these figures will be further continued in Sect. 4.5.

In this section, a simple theory describing the influence of wind flow on the characteristics of freak waves is presented. A popular model of the wind wave amplification has been developed by Miles (1957, 1996). Within the framework of the linear theory, the wave enhancement is described by the energy-transfer increment

$$\delta_w = \frac{2\beta}{\kappa^2} \frac{\rho_a}{\rho_w} \left(\frac{U^*}{C_{ph}} \right)^2, \qquad (3.30)$$

where ρ_a and ρ_w are the densities of air and water, respectively, $U^* = (c_D)^{1/2} U_w$ is the friction velocity, C_{ph} is the phase velocity of the carrier wave, U_w is the wind velocity, $c_D = 0.004$ is the drag coefficient, $\kappa = 0.4$ is the Von Kármán constant, and $\beta = 2.6$. As a result, in the framework of the linear theory, the "forced" solutions of the hydrodynamic equations can be obtained from the "free" solutions multiplied by an exponential term. Hence, the amplitude of the wave packet of Gaussian shape in deep water (3.27) becomes

$$A(X,T) = \frac{A_0}{(1+16\Omega_{env}^4 X^2/g^2)^{1/4}}$$
$$\times \exp\left(\delta_w X - \frac{\Omega_{env}^2}{1+16\Omega_{env}^4 X^2/g^2}(T-X/C_{gr})^2\right). \quad (3.31)$$

More details can be found in Touboul et al. (2008).

The amplitude of the envelope (3.31) at a fixed distance X is defined by

$$A_{\max}(X) = A_0 \left[\frac{1+16\Omega_{env}^4 X_f^2/g^2}{1+16\Omega_{env}^4(X-X_f)^2/g^2}\right]^{1/4} \exp(\delta_w X), \quad (3.32)$$

where X_f is the focusing length. For a relatively weak wind increment ($\delta_w < \Omega_{env}^2/g$), the variation of the wave amplitude along the tank is not symmetrical (in comparison to the case with no wind). The extreme value of wave amplitude is increased and achieved at point

$$X_{f,wind} = X_f + \frac{1}{4\delta_w}\left[1+\sqrt{1-\frac{g^2\delta_w^2}{\Omega_{env}^4}}\right]. \quad (3.33)$$

Figure 3.13 shows the results of comparison of the theory versus fully nonlinear simulations (Touboul et al. 2008). The case with no wind is given in Fig. 3.13a, while the dynamics with wind are presented in Fig. 3.13b. Solution (3.33) corresponds to the dashed line in Fig. 3.13b. This relationship explains the tendency of the experimental results qualitatively. But the value of the Miles increment δ_w proves to be too small, so the modification of the linear solution is insignificant to describe the experiment (see Fig. 3.13b).

The wave steepness value in the vicinity of the focusing point becomes important, and one should consider nonlinear effects due to increased steepness. Hence,

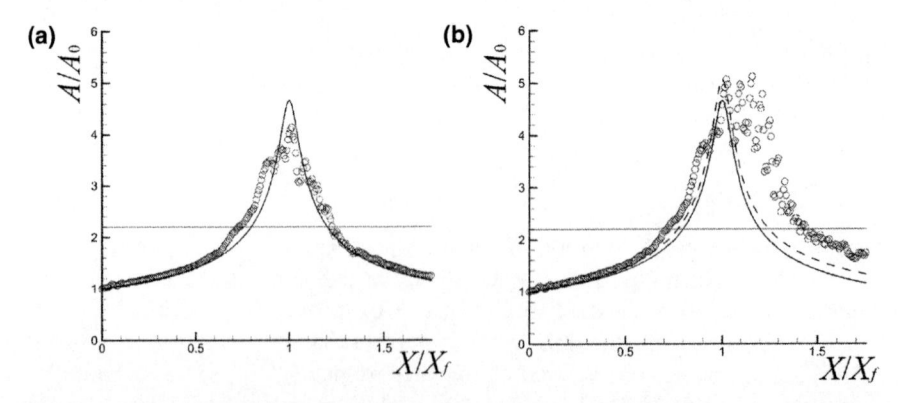

Fig. 3.13 Amplification factor as a function of normalized distance for a transient wave group: (**a**) propagated without wind (theoretical linear solution, solid line, and numerical solution, circles); (**b**) propagated under wind action with growth rate $\beta = 2.6$ (theoretical linear solution without wind, solid line, theoretical linear solution with wind (3.32), *dashed line, circles*)

the fully nonlinear hydrodynamic equations are solved within the framework of the potential nonlinear theory, as shown in the circles in Fig. 3.13a,b (more details are given in Touboul et al. (2008). Following Banner and Song (2002), the surface pressure distribution during the experiments is assumed to be of the following form:

$$P = \alpha \rho_a U^{*2} \frac{\partial \eta}{\partial X} \quad (3.34)$$

with $\alpha = 2\beta/\kappa^2$ and $U^* = 0.2 C_{ph}$. Strong nonlinearity leads to a shifting of the focused area and an increase of its width in full agreement with observed data. To demonstrate how the wind effect can modify the behavior of steep wave groups, the spatial evolution of the amplification factor A/A_0 has been computed without wind as well (see Fig. 3.13a). In this case, theoretical and numerical solutions are close, and weak deviations occur during the formation of the steep group. A nonlinear saturation in amplitude and a weak widening in the vicinity of the peak are observed. When the wind effect is introduced, the behavior of the wave train is strongly modified. This feature emphasizes the significance of the water wave nonlinearity as displayed in Fig. 3.13b.

A more careful and detailed analysis of the wind-wave interaction during the wave focusing emphasized the strong coupling between the wave group and the turbulent boundary layer when the extreme wave event occurs (Touboul et al. 2006, Kharif et al. 2008). Hence, it has been shown that air-sea fluxes are strongly enhanced in the presence of strongly nonlinear wave groups. This strong correlation between the very steep waves in the group and the wind suggests that the Jeffreys' sheltering mechanism (1925) could be a suitable model to describe this coupling. This aspect is further discussed in Sects. 4.3.4 and 4.4.

To conclude, we can claim that wind-wave interaction increases the duration and intensity of rogue wave events.

3.4 Wave-Current Interaction as a Mechanism of Rogue Waves

One of the first collections of observed rogue wave events has been gathered by Mallory (1974) for the southwestern Indian Ocean, where the Agulhas current passes along the South Africa coast (see description given in Sect. 1.1 and Fig. 1.1d). Thus, the first theoretical models of the freak wave phenomenon considered wave-current interaction (Peregrine 1976; Basovich and Talanov 1977; Thomas 1981, 1990; Lavrenov 1998, 2003; Shyu and Phillips 1990; White and Fornberg 1998; Brown 2000, 2001). Noticing that the characteristic scales of oceanic currents are large compared to wind-wave wavelengths, the ray approach described in Sect. 3.1 may be successfully applied with the dispersion relation for water waves propagating on currents. Considering the deep-water wave case, the dispersion relation for waves on a steady current becomes anisotropic for unidirectional wave propagation

$$\Omega = \Omega_i(K) + \mathbf{K} \cdot \mathbf{U_c}(X,Y), \quad \Omega_i = \pm\sqrt{gK}, \quad (3.35)$$

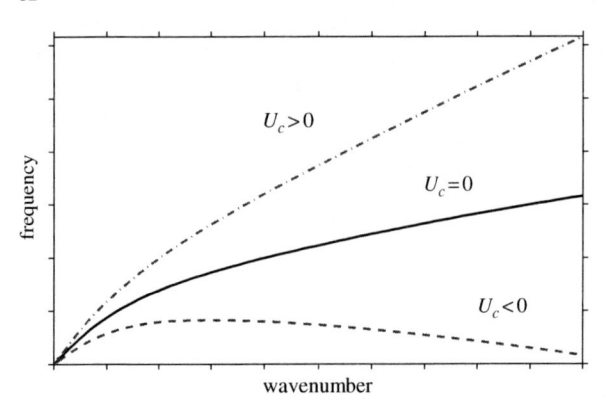

Fig. 3.14 Dispersion relation for unidirectional wave propagation over a current

Here, Ω_i is the intrinsic frequency (when there is no current). Even in the one-dimensional (1D) case, when the current passes along the OX axis and its speed is a function of one coordinate X only—$\mathbf{U}_c = (U_c(X), 0)$—the wave-current interaction is not trivial. When the current is opposite to the incident monochromatic waves, they suffer from the blocking phenomenon at the point X_*, where the group velocity (in the non-moving system of coordinates) is zero

$$C_{gr} = \frac{d\Omega}{dK} = \frac{1}{2}\sqrt{\frac{g}{K}} + U_c(X_*) = 0. \tag{3.36}$$

A wave approaching the blocking point has phase and group velocities of the same sign. After reflection from the blocking point, the group velocity gets a sign opposite to that of the phase velocity (see Fig. 3.14). The wave number increases during wave-current interaction, and an initial long wave transforms into a short wave. The wave amplitude can be found from the wave action balance equation

$$\frac{\partial}{\partial T}\left(\frac{A^2}{\Omega_i}\right) + \nabla \cdot \left(\frac{C_{gr}A^2}{\Omega_i}\right) = 0. \tag{3.37}$$

This is a generalization of the energy balance equation (3.5) for waves on current (Bretherton and Garrett 1969, Peregrine 1976). For steady currents, (3.37) results in the wave action flux-conservation law

$$C_{gr}bA^2/\Omega_i = \text{const}, \tag{3.38}$$

where b, as previously, is the distance between neighboring rays.

For unidirectional wave propagation, the blocking point characterized by zero group velocity (3.36) plays the role of a caustic where the wave amplitude formally tends to infinity. In fact, Eq. (3.37) is not valid in the vicinity of caustics, and a more accurate asymptotic analysis using the Maslov representation should be applied; it gives near caustics the following expression for the wave field (Peregrine 1976, Lavrenov 2003)

3.4 Wave-Current Interaction as a Mechanism of Rogue Waves

$$\eta(X,T) \propto \text{Ai}\left[\left(\frac{8\partial U_c/\partial X}{\Omega_i(K_*)}\right)^{1/3} K_*(X-X_*)\right]\cos(K_*X - \Omega T), \quad (3.39)$$

where $K_* = K(X_*)$ is the value of the wavenumber at the blocking point calculated from (3.36), and $\partial U_c/\partial X$ is computed at the same point X_*. As a result, the wave amplitude at the blocking point remains bounded:

$$\frac{A_*}{A_0} \propto \left(\frac{\Omega_i}{dU_c/dX}\right)^{1/6}. \quad (3.40)$$

This formula is valid for linear waves only; waves of large amplitudes are usually breaking; see photo in Fig. 3.15 of wave-blocking phenomenon at Indian River inlet (Delaware, USA) taken from Chawla and Kirby (2002). The wave blocking on opposite currents has been studied in laboratory tanks (Badulin et al. 1983, Pokazayev and Rozenberg 1983, Lai et al. 1989, Chawla and Kirby 2002). Figure 3.16 displays the process of the wave reflection from the blocking point with wavelength reduction ($X < 0$) and strong attenuation beyond this point ($X > 0$).

Transient and irregular wave groups are of special interest in the context of rogue waves. In both cases, caustics are spatially distributed and various spectral components are blocked at different points on the variable opposite current. Obtaining an analytical solution for transient groups is a difficult task and we give here only one exact solution that is valid for narrow-banded and weakly nonlinear wave groups in the vicinity of caustics[1] (Chen and Liu 1976).

Fig. 3.15 Wave blocking at Indian River inlet (USA). Reproduced from Chawla and Kirby (2002) by permission of American Geophysical Union.

---

[1] This is an exact solution of the variable-coefficient nonlinear Schrödinger equation that can be derived for such a situation.

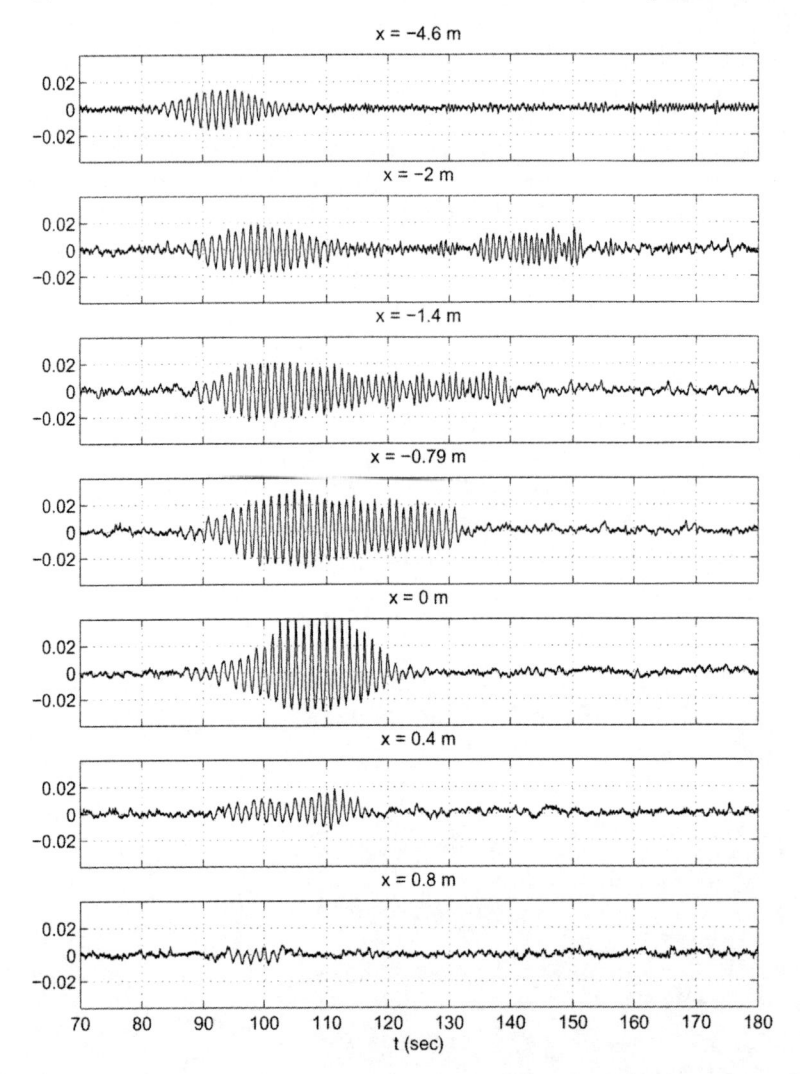

Fig. 3.16 Wave reflection and passing through blocking point in laboratory flume. Reproduced from Chawla and Kirby (2004)

$$A(X,T) = B \operatorname{sech}\left[\sqrt{2}KB(X-VT)\right] \qquad (3.41)$$
$$\times \exp\left[i\left(\sqrt{2}K^2BX + \Omega KBT^2|dU_c/dX|/\sqrt{32} - \Omega T^3|dU_c/dX|^2/24\right)\right].$$

In (3.41), Ω denotes the mean frequency, while the wavenumber K is defined at the blocking point by condition (3.36). This group may be, in fact, transformed to an envelope solution of the nonlinear Schrödinger equation; it preserves the amplitude and moves with variable speed

3.4 Wave-Current Interaction as a Mechanism of Rogue Waves

$$V = \frac{\Omega B}{\sqrt{8}} - \frac{\Omega}{4K} \left| \frac{dU_c}{dX} \right| T. \qquad (3.42)$$

The group moves towards the blocking point, penetrates it, and then is reflected. Surprisingly, the soliton amplitude B remains constant during the wave propagation in the inhomogeneous medium due to the balance between dispersive focusing (defocusing) and attenuation (amplification) in the zone of nonuniform current. This solution demonstrates that the result depends on a competition of these two effects, and therefore the amplitude of the transient group can either be amplified or attenuated at caustics in natural conditions.

When the current is uniform, and the wave maker in the flume generates a wavetrain with linearly decreasing frequency, caustics are nevertheless spatially distributed again and do not allow optimal wave-train compression. This result may be straightforwardly shown from the solution of the kinematic equation (3.12), when the Doppler effect on a constant current is taken into account (Touboul et al. 2007)

$$\Omega(X,T) = \Omega_0(\tau), \quad \tau = T - \frac{X}{C_{gr}(\Omega) + U_c}, \qquad (3.43)$$

with the condition $\partial \Omega / \partial T$ going to infinity. The focusing point is then spread over a focusing area, extending from L_{min} to L_{max}, where

$$L_{max/min} = X_f \left(1 + \frac{2U_c \Omega_{max/min}}{g} \right)^2. \qquad (3.44)$$

Irregular unidirectional wave groups have been studied experimentally and numerically by Chawla and Kirby (2002) and Wu and Yao (2004). Experimental results confirm that a random wave field does not prevent rogue-wave formation caused by dispersive focusing. Strong opposite currents inducing partial wave blocking significantly elevate the limiting steepness and asymmetry of freak waves.

Reflection of oblique waves by currents was studied analytically by Shyu and Tung (1999). A more general approach takes into account two horizontal coordinates and realistic profiles of transverse shear currents for complex ray patterns with the generation of "normal" caustics when the differential width is zero ($b = 0$), and specific "current" caustics when $C_{gr} = 0$. Lavrenov (1998, 2003) calculated the ray pattern in the vicinity of the Agulhas current for one rogue wave event, and showed that it contains focus points where the wave energy concentrates. White and Fornberg (1998) took into account the weak randomness of the current (about 5% of the wave speed) and showed that variable currents can lead to very intricate ray patterns with a large number of focal points (Fig. 3.17). The distribution of the focal points maps to a universal curve. In the ray approach, each focus corresponds to a rogue wave, but in reality the number of generated freak waves should be smaller than the number of foci. The short-lived character of rogue waves on currents can be provided by the temporal variation of the current and incident wave-front curvature.

The reported calculations demonstrate that currents can lead to the formation of rogue waves, and may be potentially met in the presence of strong currents such as

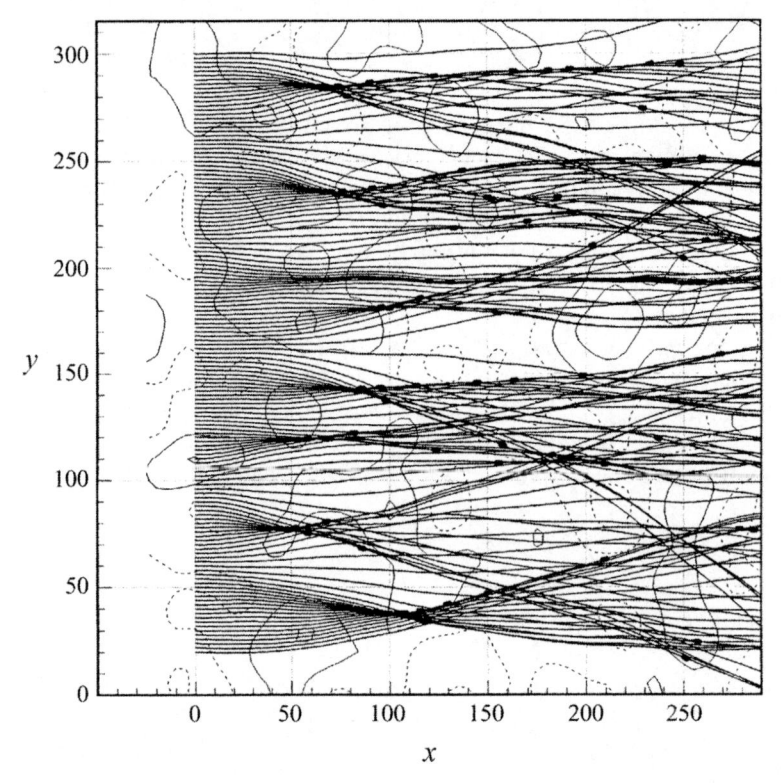

Fig. 3.17 Forming caustics due to wave-current interaction. Reproduced from White and Fornberg (1998) by permission of Cambridge University Press

the Gulf, the Agulhas, or the Kuroshio currents (Toffoli et al. 2005). The authors of the papers cited above assume that wave-current interaction is the major mechanism of the rogue wave phenomenon in deep water. In shallow water, perhaps, the wave-bottom interaction prevails.

List of Notations

A	wave amplitude
$b(X,Y)$	distance between neighbouring rays
c_D	drag coefficient
C_{gr}	group velocity
C_{LW}	long wave velocity
C_{ph}	phase velocity
D	water depth
g	acceleration due to gravity
$\mathbf{K} = (K_X, K_Y)$	wave vector
K	wavenumber

t	dimensionless time
T	time
T_f	focusing time
$\mathbf{U}_c$	current velocity
U_w	wind velocity
U^*	friction velocity
x	dimensionless coordinate
$\mathbf{X} = (X, Y)$	horizontal plane coordinate
X_f	focusing length
X_*	blocking point
δ_w	increment of energy transfer from air
$\eta(X, Y, T)$	surface elevation
κ	Von Karman constant
θ	slip angle
θ	wave complex phase
ρ_a	air density
ρ_w	water density
Ω	cyclic wave frequency
Ω_i	intrinsic frequency
∇	gradient operator

References

Abdullaev SS, Zaslavsky GM (1993) Chaos and dynamics of rays in waveguide media. Gordon & Breach Sci Publ, New York

Arnold VI (1990) Singularities of caustics and wavefronts. Kluwer Acad Publ, Dordrecht

Babic VM, Buldyrev VS (1991) Short-wavelength diffraction theory. Springer, Heidelberg

Badulin SI, Pokazayev KV, Rozenberg AD (1983) A laboratory study of the transformation of regular gravity-capillary waves in inhomogeneous flows. Izv Atmos Ocean Phys 19:782–787

Baldock TE, Swan C, Taylor PH (1996) A laboratory study of nonlinear surface waves on water. Phil Trans Roy Soc Lond A 354:649–676

Banner ML, Song J (2002) On determining the onset and strength of breaking for deep water waves. Part II: Influence of wind forcing and surface shear. J Phys Oceanogr 32:2559–2570

Basovich AYa, Talanov VI (1977) Transformation of short surface waves on inhomogeneous currents. Izv Atmos Ocean Phys 13:706–733

Bateman WJD, Swan C, Taylor PH (2001) On the efficient numerical simulation of directionally spread surface water waves. J Comput Phys 174:277–305

Bateman WJD, Swan C, Taylor PH (2003) On the calculation of the water particle kinematics arising in a directionally spread wavefield. J Comput Phys 186:70–92

Bona JL, Saut J-C (1993) Dispersive blowup of solutions of generalized Korteweg – de Vries equations. J Differ Equ 103:3–57

Bonnefoy F, de Reilhac PR, Le Touzé D, Ferrant P (2005) Numerical and physical experiments of wave focusing in short-crested seas. In: Proc. 14th Aha Huliko'a Winter Workshop, Honolulu, Hawaii, 2005

Brandini C, Grilli S (2001a) Modeling of freak wave generation in a 3D-NWT. In: Proc. 11th Int Offshore and Polar Eng Conf ISOPE , Stavanger, Norway, 2001, pp 124–131

Brandini C, Grilli S (2001b) Three-dimensional wave focusing in fully nonlinear wave models. http://www.oce.uri.edu/~grilli/focus_waves01.pdf. Accessed 18 May 2008

Brekhovskikh LM (1980) Waves in layered media. Academic Press, New York

Bretherton FP, Garrett CJR (1969) Wavetrains in inhomogeneous moving media. Proc Roy Soc Lond A 302:529–554

Brown MG (2000) The Maslov integral representation of slowly varying dispersive wavetrains in inhomogeneous moving media. Wave Motion 32:247–266

Brown MG (2001) Space-time surface gravity wave caustics: structurally stable extreme wave events. Wave Motion 33:117–143

Brown MG, Jensen A (2001) Experiments on focusing unidirectional water waves. J Geophys Res 106(C8):16917–16928

Chawla A, Kirby JT (2002) Monochromatic and random wave breaking at blocking points. J Geophys Res 107(C7):3067. doi:10.1029/2001JC001042

Chen HH, Liu CS (1976) Solitons in nonuniform media. Phys Rev Lett 37:693–697

Choi BH, Pelinovsky E, Riabov I, Hong SJ (2002) Distribution functions of tsunami wave heights. Nat Hazards 25:1–21

Clauss G (1999) Task-related wave groups for seakeeping tests or simulation of design storm waves. Appl Ocean Res 21:219–234

Clauss G (2002) Dramas of the sea: episodic waves and their impact on offshore structures. Appl Ocean Res 24:147–161

Clauss G, Bergmann J (1986) Gaussian wave packets: a new approach to seakeeping tests of ocean structures. Appl Ocean Res 8:190–206

Contento G, Codiglia R, D'Este F (2001) Nonlinear effects in 2D transient nonbreaking waves in a closed flume. Appl Ocean Res 23:3–13

Dobrokhotov SYu (1983) Maslov methods in the linearized theory of gravity waves on the surface of a liquid. Sov Phys Dokl 28:229–231

Dobrokhotov SYu, Zhevandrov PN (2003) Asymptotic expansions and the Maslov canonical operator in the linear theory of water waves. 1. Main constructions and equations for surface gravity waves. Russ J Math Phys 10:1–31

Dobrokhotov SYu, Sekerzh-Zenkovich SYa, Tirozzi B, Volkov B (2006) Explicit asymptotics for tsunami waves in framework of the piston model. Russ J Earth Sci 8:ES4003. doi: 10.2205/2006ES000215

Dysthe KB, Krogstad HE, Socquet-Juglard H, Trulsen K (2005) Freak waves, rogue waves, extreme waves and ocean wave climate. http://www.math.uio.no/~karstent/waves/ index_en.html. Accessed 14 March 2008

Draper L (1964) 'Feak' ocean waves. Oceanus 10:13–15

Engelbrecht JK, Fridman VE, Pelinovski EN (1988) Nonlinear evolution equations. Longman, London

Fochesato C, Grilli S, Dias F (2007) Numerical modeling of extreme rogue waves generated by directional energy focusing. Wave Motion 44:395–416

Goulitski K, Shemer L, Kit E (2004) Steep unidirectional waves: experiments and modeling. Izv VUZ Appl Nonlinear Dynamics 12:122–131

Jeffreys H (1925) On the formation of wave by wind. Proc Roy Soc A 107:189–206

Johannessen TB, Swan C (1997) Nonlinear transient water waves – Pt. 1. A numerical method of computation with comparisons to 2-D laboratory data. Appl Ocean Res 19:293–308

Kharif C, Giovanangeli JP, Touboul J et al (2008) Influence of wind on extreme wave events: experimental and numerical approaches. J Fluid Mech 594:209–247

Kharif C, Pelinovsky E, Talipova T, Slunyaev A (2001) Focusing of nonlinear wave groups in deep water. J Exp Theor Phys Lett 73:170–175

Lai RJ, Long SR, Huang NE (1989) Laboratory studies of wave-current interaction: kinematics of the strong interaction. J Geophys Res 94:16201–16214

Lavrenov IV (2003) Wind waves in ocean: dynamics and numerical simulations. Springer-Verlag, Heidelberg

Lavrenov I (1998) The wave energy concentration at the Agulhas current of South Africa. Nat Hazards 17:117–127

References

Magnusson AK, Donelan MA, Drennan WM (1999) On estimating extremes in an evolving wave field. Coastal Eng 36:147–163

Mallory JK (1974) Abnormal waves on the south-east of South Africa. Inst Hydrog Rev 51:89–129

Mei CC (1983) The applied dynamics of ocean surface waves. Wiley, New York

Miles JW (1957) On the generation of surface waves by shear flow. J Fluid Mech 3:185–204

Miles JW (1996) Surface-wave generation: a viscoelastic model. J Fluid Mech 322:131–145

Ostrovsky L, Potapov A (1999) Modulated waves, theory and applications. John Hopkins University Press, Baltimore

Pelinovsky E, Talipova T, Kharif C (2000) Nonlinear dispersive mechanism of the freak wave formation in shallow water. Phys D 147:83–94

Pelinovsky E, Talipova T, Kurkin A, Kharif Ch (2001) Nonlinear mechanism of the tsunami wave generation by atmospheric disturbances. Nat Hazards Earth Sys Sci 1:243–250

Pelinovsky EN (1982) Nonlinear dynamics of tsunami waves. IAP RAS Press, Nizhny Novgorod (In Russian)

Peregrine DH (1976) Interaction of water waves and currents. Adv Appl Mech 16:9–117

Peregrine DH (1983) Wave jumps and caustics in the propagation of finite-amplitude water waves. J Fluid Mech 136:435–452

Peregrine DH, Smith R (1979) Nonlinear effects upon waves near caustics. Phil Trans Roy Soc Lond A292:341–370

Pokazayev KV, Rozenberg AD (1983) Laboratory studies of regular gravity-capillary waves in currents. Oceanology 23:429–435

Shyu JH, Phillips OM (1990) The blocking of gravity and capillary waves by longer waves and currents. J Fluid Mech 217:115–141

Shyu JH, Tung CC (1999) Reflection of oblique waves by currents: analytical solutions and their application to numerical computations. J Fluid Mech 396:143–182

Slunyaev A, Kharif C, Pelinovsky E, Talipova T (2002) Nonlinear wave focusing on water of finite depth. Phys D 173:77–96

Stansberg CT (2001) Random waves in the laboratory – what is expected for the extremes? In: Olagnon M, Athanassoulis GA (eds) Rogue Waves 2000, Ifremer, France, 289–301

Thomas GP (1981) Wave-current interactions: an experimental and numerical study: Part I: linear waves. J Fluid Mech 110:457–474

Thomas GP (1990) Wave-current interactions: an experimental and numerical study: Part II: nonlinear waves. J Fluid Mech 216:505–536

Toffoli A, Lefevre JM, Bitner-Gregersen E, Monbaliu J (2005) Towards the identification of warning criteria: Analysis of a ship accident database. Appl Ocean Res 27:281–291

Touboul J, Giovanangeli JP, Kharif C, Pelinovsky E (2006) Freak waves under the action of wind: Experiments and simulations. Eur J Mech B / Fluids 25:662–676

Touboul J, Kharif C, Pelinovsky E, Giovanangeli JP (2008) On the interaction of wind and steep gravity wave groups using Miles' and Jeffreys' mechanisms. Submitted to Nonlinear Processes in Geophysics

Touboul J, Pelinovsky E, Kharif C (2007) Nonlinear focusing wave groups on current. J Korean Soc Coastal and Ocean Eng 9:222–227

White BS, Fornberg B (1998) On the chance of freak waves at the sea. J Fluid Mech 255:113–138

Whitham GB (1974) Linear and nonlinear waves. Wiley & Sons, New York London Sydney Toronto

Wu CH, Yao A (2004) Laboratory measurements of limiting freak waves on currents. J Geophys Res 109:C12002-1–18

Chapter 4
Rogue Waves in Waters of Infinite and Finite Depths

The most widely investigated rogue wave events are those due to modulational instability or dispersive focusing mechanisms. So far, the nonlinear terms of the equations have been neglected, hence in this chapter attention is paid to rogue wave occurrence when nonlinear effects are taken into account. This chapter—which is mainly devoted to modeling and simulating the physics of rogue wave events in the deep sea—addresses finite depth situations to some extent, too.

First, we present the modulational instability of water waves within the framework of the fully nonlinear equations and weakly nonlinear approximate approach in Sect. 4.1. From a deterministic viewpoint, it is the so-called Benjamin-Feir instability: a carrier wave is unstable in terms of sideband perturbations provided their respective wavenumbers are sufficiently close. From a statistical view point it is known as spectral instability, which is the random version of the Benjamin-Feir instability: a random narrowband wave train is unstable in terms of sideband perturbations provided the width of the spectrum is sufficiently narrow.

The widely-used nonlinear Schrödinger equation and related approximate theory for the Benjamin-Feir instability are presented in Sect. 4.2. Generation of rogue waves due to the nonlinear-dispersive focusing is investigated with the help of the inverse scattering approach. Breathing exact solutions of this model are described.

Section 4.3 is devoted to the occurrence of rogue waves in the deep sea when fully nonlinear equations are used. The High Order Spectral Method (HOSM) and the Boundary Integral Equation Method (BIEM), which are used to simulate numerically rogue waves due to modulational instability and dispersive focusing, are briefly presented with and without wind forcing. Sections 4.2 and 4.3 are devoted to deterministic description of the rogue-wave occurrence, while Sect. 4.4 concerns a statistical description of these giant waves.

Some laboratory experiments on rogue waves are presented in Sect. 4.5 with and without wind action.

Section 4.6 is aimed at presenting 3D aspects of the freak-wave occurrence.

Instrumental registrations of rogue waves give the possibility to fit elaborate theories with natural phenomena. Some approaches to understanding the nature of freak waves are presented in Sect. 4.7.1. They exhibit significant nonlinear (and modulational) effects when rogue waves occur. Results of statistical processing of huge wave in-situ records are collected in Sect. 4.7.2.

4.1 The Modulational Instability

The generation of extreme wave events can be simply obtained from the Benjamin-Feir instability (or modulational instability) of uniformly traveling trains of Stokes waves in water of infinite and finite depths. Stokes' wave trains are unstable in terms of various perturbations. Among these instabilities is the Benjamin-Feir instability (a long-wave instability). The latter dominates for small values of the amplitude. Various researchers discovered the existence of modulational instability of Stokes waves at the same time. Lighthill (1965) provided a geometric condition for wave instability in deep water. Later, Benjamin and Feir (1967) demonstrated the result analytically. Using a Hamiltonian approach, Zakharov (1968) derived the same instability result. Furthermore, in the context of modulated water waves, he obtained the famous Nonlinear Schrödinger equation. It would have been more appropriate to call the modulational instability the BFLZ instability instead of BF instability. Benney and Roskes (1969) extended the study to finite depth and derived what is now called the Davey-Stewartson system (Davey and Stewartson 1974). Both Zakharov (1968) and Benney and Roskes (1969), for infinite depth and finite depth, respectively, investigated the stability with 3D perturbations. Dysthe (1979) pursued the perturbation analysis one step further, to fourth-order in wave steepness, and found that the wave-induced mean flow significantly influences the growth rate of the modulational instability. Later on, Stiassnie and Shemer (1984) used a powerful approximate equation—the Zakharov equation free of the narrow band assumption—to investigate the stability of Stokes waves. Note that the Dysthe equation was derived from the Zakharov equation by Stiassnie (1984) under the assumption of narrow band wave field. Furthermore, with the Zakharov equation it is possible to consider perturbations different from modulational type. Later on, numerical computations based on fully nonlinear equations, allowed researchers to go beyond the modulational instability. The main advantage is that there is no restriction on the length of perturbations and amplitude of the basic wave. Longuet-Higgins (1978a,b) investigated 2D instabilities, whereas McLean et al. (1981) and McLean (1982a,b) considered 3D instabilities of 2D Stokes waves. More recently, Francius and Kharif (2006) extended the linear stability analysis of finite-amplitude periodic progressive gravity waves to steeper waves and shallower water. The method used by McLean to study the stability of Stokes wave trains, within the framework of the fully nonlinear equations, is presented in Sect. 4.1.1, followed by a brief presentation of the obtained main results. See the papers by Kharif and Ramamonjiarisoa (1988) and Dias and Kharif (1999), too.

4.1.1 Within the Framework of the Fully Nonlinear Equations

In this section, a general presentation of surface wave instabilities is given, based on the fully nonlinear equations (2.13), (2.28), (2.30), and (2.31). More precisely, we consider the linear stability of a Stokes' wave train of arbitrary wave steepness.

4.1 The Modulational Instability

Let $\eta = \bar{\eta} + \eta'$ and $\phi = \bar{\phi} + \phi'$ be the perturbed elevation and perturbed velocity potential, where $(\bar{\eta}, \bar{\phi})$ and (η', ϕ') correspond to the unperturbed Stokes wave (basic wave) and infinitesimal perturbative motion $(\eta' \ll \bar{\eta}, \phi' \ll \bar{\phi})$, respectively. Following Longuet-Higgins (1985), the Stokes wave of wavenumber K_0 is computed iteratively. Substituting these decompositions in the boundary conditions linearized about the unperturbed motion $(\bar{\eta}, \bar{\phi})$, and using the following forms with p and q arbitrary real numbers (see McLean 1982b),

$$\eta' = e^{-i\sigma T} \exp\left[i(pK_0 X + qK_0 Y)\right] \sum_{j=-\infty}^{\infty} A_j e^{ijK_0 X}, \tag{4.1}$$

$$\phi' = e^{-i\sigma T} \exp\left[i(pK_0 X + qK_0 Y)\right] \sum_{j=-\infty}^{\infty} B_j e^{ijK_0 X} \frac{\cosh\left[\sqrt{(p+j)^2 + q^2} K_0 (Z+D)\right]}{\cosh\left[\sqrt{(p+j)^2 + q^2} K_0 D\right]}, \tag{4.2}$$

yields a complex eigenvalue problem for σ, with eigenvector $\mathbf{u} = (A_j, B_j)^t$

$$(\mathbf{A} - i\sigma \mathbf{B})\mathbf{u} = 0, \tag{4.3}$$

where $\mathbf{A}$ and $\mathbf{B}$ are complex matrices depending on the wave steepness of the basic wave and the arbitrary real numbers pK_0 and qK_0 corresponding to the longitudinal and transverse wavenumbers of the perturbation, respectively. The physical disturbance that corresponds to the real part of expressions (4.1) and (4.2) has periods $2\pi/pK_0$ and $2\pi/qK_0$ in X- and Y-directions, respectively. The terms corresponding to the sums in (4.1) and (4.2) have the spatial periodicity of the basic Stokes waves. Hence, forms (4.1) and (4.2) express that the perturbations feel the presence of the Stokes waves. Instability corresponds to $\mathrm{Im}(\sigma) \neq 0$. The spectrum is easy to compute when $\bar{\eta}(X, T) = 0$. In the moving frame with the basic wave, one finds that the eigenvalues are

$$\begin{aligned} \sigma_n &= -(p+n)\sqrt{gK_0 \tanh(K_0 D)} \pm \sqrt{gK_n \tanh(K_n D)}, \\ k_n^2 &= (p+n)^2 + q^2, \quad K_n = K_0 k_n. \end{aligned} \tag{4.4}$$

The eigenvalues are real, hence the state corresponding to $\bar{\eta} = 0$ is spectrally stable. As the wave steepness of the Stokes wave increases, the eigenvalues move. MacKay and Saffman (1986) derived a necessary condition for a Stokes wave to lose spectral stability corresponding to the collision of eigenvalues of opposite Krein signature (Krein 1955), or a collision of eigenvalues at zero (see MacKay and Saffman 1986).

$$\sigma_{n_1}^{\pm}(p,q) = \sigma_{n_2}^{\pm}(p,q) \tag{4.5}$$

The instabilities are separated into two classes: class I when the collisions occur between modes with $n = m$ and $n = -m$, and class II when the collisions occur between modes with $n = m$ and $n = -m - 1$. The corresponding instabilities are called class I and class II instabilities. Class I (m) corresponds to ($2m+2$)-wave interactions, whereas class II (m) corresponds to ($2m+3$)-wave interactions:

Class I (m)

$$\mathbf{k}_1 = (m+p,q), \quad \mathbf{k}_2 = (m-p,-q),$$
$$\sigma_m^+(p,q) = \sigma_{-m}^-(p,q), \qquad (4.6)$$
$$\Omega_1 + \Omega_2 = 2m\Omega_0.$$

Class II (m)

$$\mathbf{k}_1 = (m+p,q), \quad \mathbf{k}_2 = (1+m-p,-q),$$
$$\sigma_m^+(p,q) = \sigma_{-m-1}^-(p,q), \qquad (4.7)$$
$$\Omega_1 + \Omega_2 = (2m+1)\Omega_0,$$

with

$$\Omega_n = \sqrt{gK_n \tanh(K_n D)}, \quad n=0,1,2,$$
$$\mathbf{K}_n = K_0 \mathbf{k}_n, \quad K_n = K_0 k_n, \quad n=1,2. \qquad (4.8)$$

The collision of eigenvalues may be interpreted as wave-wave resonant interactions satisfying the following conditions

$$\mathbf{K}_1 + \mathbf{K}_2 = N\mathbf{K}_0, \quad \Omega_1 + \Omega_2 = N\Omega_0, \quad N \geq 2, \qquad (4.9)$$

where even values of $N(=2m)$ correspond to Class I (m), and odd values of $N(=2m+1)$ correspond to Class II (m), respectively.

Class I (m) instabilities correspond to resonant interactions between the basic mode $\mathbf{K}_0 = (1,0)K_0$ counted $2m$ times and the satellites $\mathbf{K}_1 = (m+p,q)K_0$ and $\mathbf{K}_2 = (m-p,-q)K_0$, whereas class II ($m$) instabilities correspond to resonant interactions between the basic mode $\mathbf{K}_0 = (1,0)K_0$ counted $2m+1$ times and two satellites $\mathbf{K}_1 = (m+p,q)K_0$, $\mathbf{K}_2 = (1+m-p,-q)K_0$. For instance, $N=2$ corresponds to quartet resonant interactions, and $N=3$ responds to quintet resonant interactions, etc.

The BF instability belongs to class I instability with $m=1$ and corresponds to small values of the wavenumber p. Class I ($m=1$) generalizes the BF instability and includes modulational instabilities.

In water of infinite depth ($K_0 D \to \infty$), the 2D ($q=0$) modulational instability is dominant for small to moderate values of the wave steepness, whereas for larger values, 3D instabilities of class II ($m=1$) become dominant. The latter instability may lead to the formation of horseshoe patterns while modulational instability evolves into a series of modulation-demodulation cycles (Fermi-Pasta-Ulam recurrence).

In finite depth, McLean (1982b) considered three depths—one greater ($K_0 D = 2$) and two smaller ($K_0 D = 1$ and 0.5) than $K_0 D = 1.363$, which is a critical depth (see the next section). He confirmed the stabilization of 2D long-wave perturbations ($p \ll 1$) for $K_0 D < 1.363$ as predicted by Whitham (1967). For $K_0 D = 2$, he found that the dominant instability still belongs to class I ($m = 1$) when the wave steepness is small or moderate. Unlike the deep water case, the modulational instability is now 3D ($q \neq 0$). For steeper waves, 3D instabilities of class II ($m = 1$) crescent-shaped form become dominant. For $K_0 D = 1$, 2D long-wave perturbations ($p \ll 1$) of class I ($m = 1$) are stable for small wave steepness. However, this class is dominated by 3D unstable perturbations ($q \neq 0$). For steeper waves, it is the crescent-shaped instability of class II ($m = 1$) that is dominant. The shallowest case ($K_0 D = 0.5$) that McLean considered is most unstable for small wave steepness, to a 2D perturbation of class I ($m = 1$) with a wavenumber comparable to K_0, in contrast to the familiar 2D long-wave perturbations that are the dominant instabilities in deep water. For small-amplitude waves, this result was rediscovered by Francius and Kharif (2006) for $K_0 D < 0.5$. Two-dimensional long-wave perturbations are stable at these depths. For moderate steepness, the dominant instability shifts to the 3D one and is still associated with class I ($m = 1$). For sufficiently steep waves, class II ($m = 1$) dominates and the most unstable perturbation is three-dimensional.

4.1.2 Within the Framework of the Nonlinear Schrödinger (NLS) Equation

The evolution equations describing wave propagation over deep or shallow waters may straightforwardly be derived heuristically (Kharif and Pelinovsky 2006). One of the common ways to rigorously derive these equations is based on the asymptotic technique of Engelbrecht et al. (1988). Slowly modulated weakly nonlinear water waves may thus be described with the help of approximate asymptotic equations for wave modulations. The Nonlinear Schrödinger (NLS) equation represents the simplest equation of this kind, first obtained by Zakharov (1968). The detail of its derivation may be found, for example, in Johnson (1997).

4.1.2.1 The Davey-Stewartson and Nonlinear Schrödinger Equations

Let us consider unidirectional wave propagation on the sea surface of arbitrary constant depth; the geometry of the problem is the same as used in Chap. 2 (see Fig. 2.1). The system of governing equations is given by the Laplace equation (2.13), boundary conditions on the free surface (2.28) and (2.30), and the sea bottom condition (2.46).

We will restrict our interest to the narrow-band wave fields (long-wave modulations) so that the solution to the problem may be sought in the form of perturbation expansions similar to (2.33) and (2.34):

$$\phi(X,Y,Z,T) = \sum_{n=0}^{\infty} \varepsilon^{n+1} \phi_n(X,Y,Z,T), \tag{4.10}$$

$$\eta(X,Y,Z,T) = \sum_{n=0}^{\infty} \varepsilon^{n+1} \eta_n(X,Y,Z,T), \tag{4.11}$$

where

$$\phi_n = \sum_{m=-\infty}^{\infty} \phi_{n,m} E^m, \qquad (4.12)$$

$$\eta_n = \sum_{m=-\infty}^{\infty} \eta_{n,m} E^m, \qquad (4.13)$$

$$E^m = \begin{cases} 1, & m = 0 \\ 1/2 \exp\left[im\left(K_0 X - \Omega_0 T\right)\right], & m \neq 0 \end{cases}.$$

For the sake of simplicity, we choose the direction of wave propagation along the OX axis so that the carrier (fundamental) wave has wave vector $\mathbf{K}_0 = (K_0, 0), K_0 > 0$, and cyclic frequency $\Omega_0 > 0$; ε is a small parameter that will be specified later. Relations $\phi_{n,-m} = \phi_{n,m}^*$ and $\eta_{n,-m} = \eta_{n,m}^*$ should be satisfied to provide real values of the surface displacement and the velocity potential. The asterisk denotes the complex conjugate.

With the help of Taylor expansion at the still water level (2.35), the boundary conditions on the sea surface read

$$\eta_T + \eta_X \sum_{j=0}^{\infty} \frac{\eta^j \partial_Z^j \phi_X}{j!} + \eta_Y \sum_{j=0}^{\infty} \frac{\eta^j \partial_Z^j \phi_Y}{j!} - \phi \sum_{j=0}^{\infty} \frac{\eta^j \partial_Z^{j+1} \phi}{j!} = 0 \text{ on } Z = 0, \quad (4.14)$$

$$\sum_{j=0}^{\infty} \frac{\eta^j \partial_Z^j \phi_T}{j!} + \frac{1}{2}\left(\sum_{j=0}^{\infty} \frac{\eta^j \partial_Z^j \phi_X}{j!}\right)^2 + \frac{1}{2}\left(\sum_{j=0}^{\infty} \frac{\eta^j \partial_Z^j \phi_Y}{j!}\right)^2 + \frac{1}{2}\left(\sum_{j=0}^{\infty} \frac{\eta^j \partial_Z^{j+1} \phi}{j!}\right)^2$$
$$+ g\eta = 0 \text{ on } Z = 0. \qquad (4.15)$$

We introduce slow coordinates X_1 and Y_1, and multiple slow times T_1 and T_2 as

$$\frac{\partial}{\partial X} \Rightarrow \frac{\partial}{\partial X_0} + \varepsilon \frac{\partial}{\partial X_1}, \qquad (4.16)$$

$$\frac{\partial}{\partial Y} \Rightarrow \frac{\partial}{\partial Y_0} + \varepsilon \frac{\partial}{\partial Y_1}, \qquad (4.17)$$

$$\frac{\partial}{\partial T} \Rightarrow \frac{\partial}{\partial T_0} + \varepsilon \frac{\partial}{\partial T_1} + \varepsilon^2 \frac{\partial}{\partial T_2} + \ldots. \qquad (4.18)$$

The main contribution in the series (4.12) and (4.13) corresponds to the first harmonic ($m = \pm 1$), so that we put $\varphi_{0,m} = 0$ for $|m| > 1$ and $\eta_{0,m} = 0$ for $m \neq \pm 1$. The term $\varphi_{0,0}$ is responsible for the nonlinear induced flow (see Johnson 1997), and is also a zero-order term. Substituting series (4.16), (4.17), and (4.18) into Eqs. (4.14) and (4.15), and collecting terms of similar harmonic component (power of E) and of similar order (power of ε), one comes to a set of equations that may be resolved.

In particular, the Laplace equation (2.13) results in

$$L_m \phi_{n,m} + 2imK_0 \partial_{X_1} \phi_{n-1,m} + \partial_{X_1}^2 \phi_{n-2,m} = 0, \quad L_m = \partial_Z^2 - m^2 K_0^2. \qquad (4.19)$$

4.1 The Modulational Instability

The consecutive order-by-order solution of Eq. (4.19) provides the modal (vertical) structure of the surface waves. The leading order mode of the 2D carrier wave is given by (2.47).

Terms with ($n = 0$, $m = 1$) give the dispersion relation (2.52) and the relation between the surface disturbance and the velocity potential:

$$\eta_{0,1} = i\frac{\Omega_0}{g}\phi_{0,1}. \tag{4.20}$$

This relation was also obtained in Chap. 2 (see Eqs. (2.47) and (2.48)). The next order ($n = 1$, $m = 1$) leads to the equation

$$\frac{\partial \eta_{0,1}}{\partial T_1} + C_{gr}\frac{\partial \eta_{0,1}}{\partial X_1} = 0, \tag{4.21}$$

where C_{gr} is the group velocity (2.54), which is given by

$$C_{gr} = \frac{\partial \Omega}{\partial K} = \frac{g}{2\Omega_0}\left[\tilde{d} + K_0 D\left(1 - \tilde{d}^2\right)\right], \quad \tilde{d} \equiv \tanh(K_0 D). \tag{4.22}$$

To obtain the next order evolution equation, the neighboring harmonic components should be considered ($m = 0, 1, 2$). So, the zeroth and the second harmonics contribute to the carrier wave at this level of accuracy. These orders, solved jointly, give the following equations (Johnson 1997):

$$-i\frac{\partial \eta_{0,1}}{\partial T_2} + \beta_{11}\frac{\partial^2 \eta_{0,1}}{\partial X_1^2} + \beta_{22}\frac{\partial^2 \eta_{0,1}}{\partial Y_1^2} + \alpha_{11}|\eta_{0,1}|^2\eta_{0,1} + \alpha_{12}\eta_{0,1}\frac{\partial \phi_{0,0}}{\partial X_1} = 0, \tag{4.23}$$

$$s_1\frac{\partial^2 \phi_{0,0}}{\partial X_1^2} + s_2\frac{\partial^2 \phi_{0,0}}{\partial Y_1^2} = \gamma\frac{\partial |\eta_{0,1}|^2}{\partial X_1}. \tag{4.24}$$

The summation of Eqs. (4.21) and (4.23), supplemented by (4.24), gives the closed system of equations involving terms of two orders of accuracy:

$$-i\left(\frac{\partial A}{\partial T} + C_{gr}\frac{\partial A}{\partial X}\right) + \beta_{11}\frac{\partial^2 A}{\partial X^2} + \beta_{22}\frac{\partial^2 A}{\partial Y^2} + \alpha_{11}|A|^2 A + \alpha_{12}A\frac{\partial B}{\partial X} = 0, \tag{4.25}$$

$$s_1\frac{\partial^2 B}{\partial X^2} + s_2\frac{\partial^2 B}{\partial Y^2} = \Gamma\frac{\partial |A|^2}{\partial X}, \tag{4.26}$$

where $A \equiv \eta_{0,1}$ and $B \equiv \phi_{0,1}$. The small parameter ε used for the derivation of the model, actually defines two small quantities. They are the wave steepness (used when writing the Taylor expansions (4.14) and (4.15)) and spectral bandwidth (see (4.16), (4.17), and the series (4.12) and (4.13)). In the present approach, these quantities are supposed to be of the same order of smallness. The field of the surface displacement and velocity potential are defined according to (4.12), (4.13), and (4.20) as

$$\eta = \text{Re}\left(A\exp\left[i(K_0 X - \Omega_0 T)\right]\right), \tag{4.27}$$

$$\phi = \frac{g}{\Omega_0}\,\text{Im}\left(A\exp\left[i(K_0 X - \Omega_0 T)\right]\right). \tag{4.28}$$

The systems (4.25) and (4.26) were found by Benney and Roskes (1969) and Davey and Stewartson (1974) and are usually referred to as the Davey-Stewartson system or equation (DS). The two first terms in the LHS of Eq. (4.25) support wave propagation with linear group velocity. The four first terms in the LHS of Eq. (4.25) represent the linear dispersive part. Besides the strict asymptotic calculations, the linear dispersive contribution may easily be obtained heuristically from the dispersion relation (2.57) by using a Taylor expansion about the wave vector of the carrier, $\mathbf{K}_0 = (K_0, 0)$

$$\Omega(K_0 + K_X, K_Y) \approx \Omega(K_0, 0)$$
$$+ K_X \left.\frac{\partial \Omega}{\partial K_X}\right|_{(K_0,0)} + \frac{1}{2}K_X^2 \left.\frac{\partial^2 \Omega}{\partial K_X^2}\right|_{(K_0,0)} + \frac{1}{2}K_Y^2 \left.\frac{\partial^2 \Omega}{\partial K_Y^2}\right|_{(K_0,0)}, \tag{4.29}$$

where the derivatives give the coefficients of the linear part of Eq. (4.25),

$$C_{gr} = \frac{\partial \Omega}{\partial K}, \quad \beta_{11} = -\frac{1}{2}\frac{\partial^2 \Omega}{\partial K^2}, \quad \beta_{22} = -\frac{C_{gr}}{2K_0}. \tag{4.30}$$

The group velocity in (4.30) is given by (4.22). It may be easily seen from Fig. 2.3 that the second derivative of the frequency with respect to the wave number (which is equal to the derivative $C_{gr}'(K)$) is negative for all depths, and therefore the coefficient β_{11} is always positive, whereas β_{22} is negative.

The other coefficients in the DS system (4.25) and (4.26) are

$$\alpha_{11} = \frac{g^2 \Omega_0}{16 C_{ph}^4}\left(1 + 9\tilde{d}^{-2} - 13\left(1 - \tilde{d}^2\right) - 2\tilde{d}^4\right),$$

$$\alpha_{12} = \frac{\Omega_0}{2C_{ph}^2}\left(2C_{ph} + C_{gr}\left(1 - \tilde{d}^2\right)\right), \tag{4.31}$$

$$s_1 = C_{LW}^2 - C_{gr}^2, \quad s_2 = C_{LW}^2, \quad \Gamma = -\alpha_{12}\frac{g^2}{2\Omega_0}.$$

The long-wave speed C_{LW} in (4.31) is defined as

$$C_{LW} = \sqrt{gD}. \tag{4.32}$$

Coefficients s_1, s_2, α_{11}, and α_{12} are always positive.

The nonlinear part in Eq. (4.25) includes the effect of nonlinear induced flow (the Stokes flow), described by Eq. (4.26). In the deep-water limit, $s_1 \to \infty$ and $s_2 \to \infty$; therefore, the contribution of the mean flow velocity potential B vanishes, and (4.25) and (4.26) become

4.1 The Modulational Instability

$$-i\left(\frac{\partial A}{\partial T}+C_{gr}\frac{\partial A}{\partial X}\right)+\frac{\Omega_0}{8K_0^2}\frac{\partial^2 A}{\partial X^2}-\frac{\Omega_0}{4K_0^2}\frac{\partial^2 A}{\partial Y^2}+\frac{\Omega_0 K_0^2}{2}|A|^2 A=0, \quad (4.33)$$

where

$$\Omega_0=\sqrt{gK_0}, \quad C_{gr}=\frac{\Omega_0}{2K_0}. \quad (4.34)$$

Eq. (4.33) is the 2D (2D+1) NLS equation valid for the case of deep water (infinite depth).

The DS system (4.25) and (4.26) transforms into one evolution equation describing waves in the OXZ plane when the transverse dynamic is disregarded. Then, the DS equation results in the 1D NLS equation (Zakharov 1968, Hasimoto and Ono 1972), written as follows

$$-i\left(\frac{\partial A}{\partial T}+C_{gr}\frac{\partial A}{\partial X}\right)+\beta\frac{\partial^2 A}{\partial X^2}+\alpha|A|^2 A=0, \quad (4.35)$$

$$\beta=\beta_{11}, \quad \alpha=\alpha_{11}+\alpha_{12}\frac{\Gamma}{s_1}. \quad (4.36)$$

Contrary to the deep-water limit, the term of induced flow becomes very important in shallow water, although on deep water $\alpha > 0$, it becomes negative when the normalized depth of the basin is less than a critical value $KD = 1.363$. This bifurcation value corresponds to a significant change in the nonlinear wave dynamics. For $KD \approx 1.363$, the nonlinear coefficient in Eq. (4.35) turns to zero, and thus, the nonlinear effects appear at higher levels, and may be taken into account through a modified asymptotic scheme (see Johnson 1977, Kakutani and Michihiro 1983, Sedletsky 2003, and Slunyaev 2005).

To conclude this section, we would like to state here two important remarks about the NLS equation. First, the coefficients in the evolution equations turn out to be functions of the carrier wave frequency Ω_0 (or wavenumber K_0). The meaning of this result is illustrated by the expansion (4.29). When deriving the NLS equation, the linear dispersion relation $\Omega(K)$ is approximated by a parabolic function in the vicinity of the carrier wavenumber. Hence, to derive the DS system or NLS-like equation, it is first necessary to define the mean frequency (or wavenumber) of the waves. Although some regular methods of the mean frequency definition exist—for instance, via the spectral moments (see Sect. 2.2.2)—the result is not always robust if the waves are not sufficiently narrow-band. Secondly, the derivation of Eqs. (4.25) and (4.26) supposes two weak effects: (i) nonlinearity, of which the smallness serves for expansions (4.10), (4.11) and the expressions of the boundary conditions in the Taylor expansions (4.14) and (4.15); and (ii) weak modulation (the narrowband approximation) that is employed when considering different harmonics and introducing the slow coordinates (4.16), (4.17). In the derived equations, it is assumed that these effects are of the same order of strength. Otherwise, it is necessary to include additional terms in the evolution equation (Trulsen 2006).

4.1.2.2 The Benjamin-Feir Instability

The Benjamin-Feir instability discussed in the previous section can be studied within the framework of the NLS equation, too. Let us consider the linear stability of a plane wave of real constant amplitude A_0, frequency Ω, and wavenumber K. The solution of the NLS equation (4.35) is sought in the following form

$$A(X,T) = A_0(1+a)\exp[i(KX - \Omega T)], \tag{4.37}$$

where a is a complex function of X and T, so that $|a| \ll A_0$, and $A(X,T)$ is the exact solution of (4.35) when $a \equiv 0$ (which implies that $\Omega = C_{gr}K - \beta K^2 + \alpha A_0^2$). Then, the wave modulation a may exponentially grow with time when the following condition is satisfied:

$$\alpha\beta > 0. \tag{4.38}$$

This classical result can be found in Newell (1981), Johnson (1997), and Dias and Kharif (1999) and is true when $K_0 D > 1.363$. The long perturbations of wavenumber ΔK satisfying

$$0 < \Delta K < \Delta K_{BF}, \quad \Delta K_{BF} = A_0\sqrt{\frac{2\alpha}{\beta}} \tag{4.39}$$

are unstable, while the growth rate is given by

$$\sigma_{BF} = |\Delta K|\sqrt{2\alpha\beta A_0^2 - \beta^2 \Delta K^2}, \tag{4.40}$$

where $\sigma_{BF} \equiv -\text{Im}(\sigma)$. The subscript "BF" refers to the Benjamin-Feir instability. The maximum growth rate

$$\sigma_{BF\,\text{max}} = \alpha A_0^2 \tag{4.41}$$

is achieved for wavenumber

$$\Delta K_{BF\,\text{max}} = A_0\sqrt{\frac{\alpha}{\beta}}. \tag{4.42}$$

Let us consider a plane wave with carrier wave vector $\mathbf{K}_0 = (K_0, 0)$, perturbed by a disturbance of wave vector $\Delta \mathbf{K} = (\Delta K_X, \Delta K_Y)$. It is convenient to deal with the deep-water 2D NLS equation (4.33) (or, similarly, the DS system when the constant water case is considered) along the perturbation direction. Hence, the analysis is similar to the 1D case and results in formulas (4.38), (4.39), (4.40), (4.41) and (4.42), where the coefficients should be chosen as

$$\alpha = \frac{\Omega_0 K_0^2}{2}, \quad \beta = \frac{\Omega_0}{8K_0^2}\frac{\Delta K_X^2 - 2\Delta K_Y^2}{\Delta K_X^2 + \Delta K_Y^2}, \tag{4.43}$$

and $\Delta K_X^2 + \Delta K_Y^2 = \Delta K^2$. Instability occurs for long wave perturbations with wave vectors lying in an angular domain bounded by angles $\pm \text{atan}(2^{-1/2})$ (it is about $\pm 35°$).

4.1 The Modulational Instability

The analysis for the DS system (4.25) and (4.26) is trickier, but still may be completed analytically (Slunyaev et al. 2002). The DS system for weak 2D modulation of type (4.37) may be reduced to the form (4.35) with coefficients

$$\alpha = \left[\alpha_{11} + \alpha_{12}\frac{\Gamma}{s_1} + \alpha_{11}\frac{s_2}{s_1}\left(\frac{\Delta K_Y}{\Delta K_X}\right)^2\right] \cdot \left[1 + \frac{s_2}{s_1}\left(\frac{\Delta K_Y}{\Delta K_X}\right)^2\right]^{-1},$$

$$\beta = \left[\beta_{11} + \beta_{22}\left(\frac{\Delta K_Y}{\Delta K_X}\right)^2\right] \cdot \left[1 + \left(\frac{\Delta K_Y}{\Delta K_X}\right)^2\right]^{-1}, \quad (4.44)$$

which should be substituted into Eqs. (4.38), (4.39), (4.40), (4.41), and (4.42) to derive the stability analysis.

The instability diagram depends on the water depth and direction of propagation of the perturbation with wave vector $\Delta \mathbf{K} = (\Delta K_X, \Delta K_Y)$, which is defined by the tangent of $\Delta K_Y / \Delta K_X$. The instability diagrams in the $(\Delta K_X, \Delta K_Y)$-plane are given in Fig. 4.1 for various depths. The value of the growth rate σ_{BF} varies from zero (black) to maximum (white). For $K_0 D < 1.363$, the longitudinal perturbations

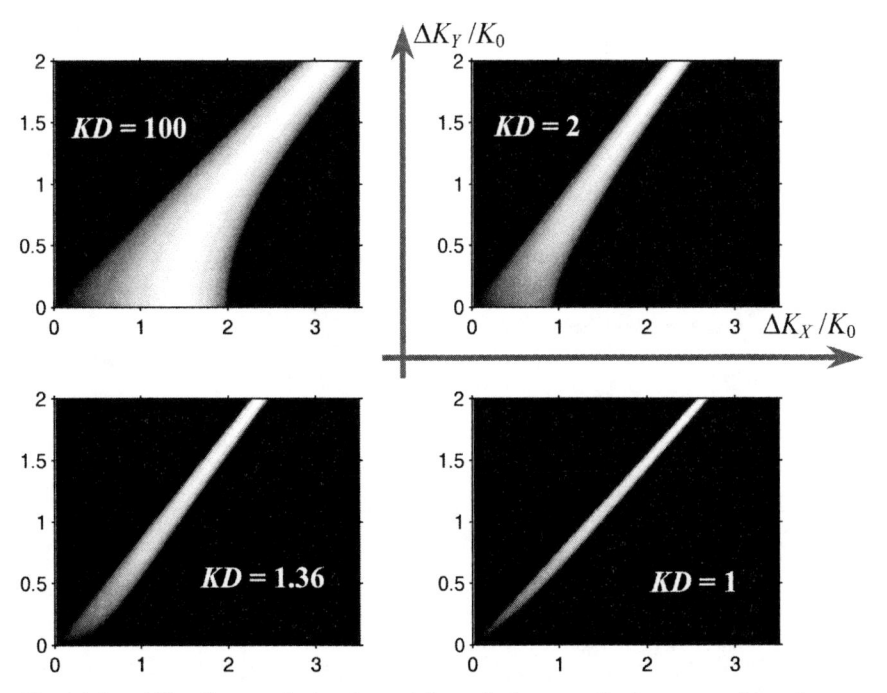

Fig. 4.1 Instability diagrams in the plane of dimensionless perturbation vectors for a plane wave with amplitude $A_0 = 1$ within the framework of the Davey-Stewartson system. Four dimensionless water depths are considered as examples: $K_0 D = 100$ (deep water), $K_0 D = 2$ (moderately deep water), $K_0 D = 1.363$ (cancellation of the nonlinear term in the evolution equation), and $K_0 D = 1$ (finite depth close to shallow water)

become stable. Only oblique perturbations develop modulational instability. For $K_0 D < 0.5$, the region of instability becomes very narrow, and the instability does not exist practically.

Formulas (4.38), (4.39), (4.40), (4.41), (4.42), (4.43), and (4.44) are valid for the weakly nonlinear theory and are based on a narrowband wave-field approximation. That is why the instability diagram has to be improved by using higher-order models (Trulsen et al. 2000).

The figure shows stability diagrams for various depths from $K_0 D = 100$ (deep water) to $K_0 D = 1$ (finite depth). As it has been already noted, only oblique perturbations suffer from modulational instability when $K_0 D < 1.363$. Generally, the instability regions become smaller when improved models are considered (see Trulsen et al. 2000). The nonlinear stage of BF instability was thoroughly investigated analytically, numerically, and experimentally (see Dias and Kharif 1999).

The Benjamin-Feir instability is one of several other possible unstable wave configurations—i.e., weak perturbations of a uniform plane wave. Other wave systems and structures may be analyzed with respect to stability (nonlinear wave packets, short-crested and bound waves), as reviewed in papers by Roskes (1976), Dhar and Das (1991), Shukla et al. (2006), and Onorato et al. (2006a). This analysis is trickier technically, less evident, and needs further research. Moreover, only the *linear* stability analysis was performed, but nonlinear instabilities are possible as well.

The nonlinear coefficient α for unidirectional waves described by (4.35) vanishes at depth $K_0 D = 1.363$ and becomes negative in shallower water; the longitudinal perturbations become stable. Oblique perturbations remain unstable, although the areas of instability shrink. This degeneration of the coefficient due to the specific geometry changes the parity of nonlinear and dispersive terms and requires consideration of higher-order asymptotic expansions, briefly considered just above, to have explicit nonlinear terms included in the evolution equation. The corresponding equation was first derived by Johnson (1977). The equation has the form

$$-i\left(\frac{\partial A}{\partial T} + C_{gr}\frac{\partial A}{\partial X}\right) + \beta\frac{\partial^2 A}{\partial X^2} + \alpha |A|^2 A - i\gamma_1 |A|^2 \frac{\partial A}{\partial X} - i\gamma_2 A^2 \frac{\partial A^*}{\partial X} + \alpha_2 |A|^4 A = 0, \tag{4.45}$$

where γ_1 and γ_2 relate to the nonlinear-dispersion contribution, and α_2 is the nonlinear coefficient of a higher (fifth) order. Considering the plane wave solution with amplitude A_0 and wavenumber K_0, the condition of possible BF instability excitation is modified when compared with (4.38) and is

$$\beta\alpha + \beta(K - K_0)(\gamma_2 - \gamma_1) + A_0^2\left(2\beta\alpha_2 - \frac{1}{2}\gamma_2^2\right) > 0. \tag{4.46}$$

Hence, the instability diagram for this marginal case is affected by the wave amplitude and the frequency offset, and qualitatively depends on the combination of coefficients in (4.46). The coefficients were first obtained by Johnson (1977), later by Kakutani and Michihiro (1983) and Slunyaev (2005), and partly by Sedletsky (2003). The complexity of the computation of high-order asymptotic expansions results in a difference of the coefficients, so that the analysis of the modulational instability in the cited papers differs quantitatively, and even sometimes qualitatively. Johnson (1977) concludes that the marginal depth when the BF instability disappears is even larger than $K_0 D \approx 1.363$, owing to the nonlinear corrections (the last summand on the LHS of (4.46)). Results obtained by Kakutani and Michihiro (1983) and Slunyaev (2005) point at the opposite conclusion: the marginal depth becomes shallower. Sedletsky (2006) undertook a further theoretical study of modulational instability within an improved generalized envelope equation theory.

4.1.2.3 The Spectral Instability of Benjamin-Feir Type

In the real sea, the wave field always suffers from random disturbances, which calls statistical considerations. Alber and Saffman (1978) and Alber (1978) derived an equation describing the evolution of the wave envelope of a random wave train. Their analysis started from the DS system, and resulted in a transport equation (see Chap. 2). Using a more general approximate equation, the Zakharov equation (Crawford et al. 1980) investigated the evolution of a random inhomogeneous field of nonlinear deep-water gravity waves. Following Alber and Saffman (1978), they considered the stability of a narrow-band homogeneous spectrum to inhomogeneous perturbations in the limiting cases of the 1D and 2D NLS equations. Using a more realistic spectrum, they obtained results that agree qualitatively with those of Alber and Saffman—namely, they found that the effect of randomness characterized by the spectral bandwidth is to reduce the growth rate and the extent of the instability.

Let us stay now within the framework of the deep-water limit of the 1D version of the NLS equation (4.33). The instability growth rate in the presence of randomness is given by Alber (1978) by

$$\frac{\sigma_{BF}}{\Omega_0} = \frac{1}{8}\frac{\Delta K}{K_0}\left[\sqrt{16(K_0 A_{rms})^2 - \left(\frac{\Delta K}{K_0}\right)^2} - \frac{2\sigma_r}{K_0}\right], \tag{4.47}$$

when random waves are distributed according to the Gaussian function and σ_r^2 is the variance that characterizes randomness effects. Variable A_{rms} denotes the root mean-square wave amplitude of the Gaussian random process (if one identifies $A_0 = 2^{1/2} A_{rms}$, then in the limit $\sigma_r \to 0$ (4.47) coincides with (4.39)). The waves are stable with respect to the BF instability if

$$\frac{\sigma_r}{K_0} > 2A_{rms}K_0. \tag{4.48}$$

In general, the effect of increasing randomness is to restrict the instability criterion, to delay the onset of instability, and to reduce the amplification rate of the modulation. The correlation length scale in the system is defined by σ_r^{-1}, and hence decorrelation of the waves (small correlation length or large σ_r) leads to stabilization of the wavetrain according to the relation

$$\frac{modulation\ length}{correlation\ length} \propto \frac{\sigma_r}{\Delta K_{BF\,max}} \propto \frac{\sigma_r K_0}{A_{rms} K_0}. \quad (4.49)$$

In fact, Alber (1978) estimated that the typically measured sea wave parameters result in stable wave trains, although they are close to the neutral stability condition.

When breaking is neglected, wave damping is usually not taken into account considering sea wave dynamics. Nevertheless, loss of energy is always observed in experiments and motivates researchers to argue whether the BF instability is relevant for real waves in the ocean. Generalizations of the NLS equation have been suggested to take into account the effects of wave dissipation and bottom friction in a simple way. Dissipative effects can be introduced in the NLS equation through a linear term with coefficient δ_{dis}

$$-i\left(\frac{\partial A}{\partial T} + C_{gr}\frac{\partial A}{\partial X}\right) + \beta\frac{\partial^2 A}{\partial X^2} + \alpha|A|^2 A - i\delta_{dis}A - i\delta_{fric}|A|^\gamma A = 0. \quad (4.50)$$

Voronovich et al. (2008) considered the effect of bottom friction that brought a more sophisticated term into the NLS equation (with coefficient δ_{fric} in (4.50)).

In Eq. (4.50), the parameter $\delta_{dis} \geq 0$ characterizes the effect of linear dissipation. δ_{fric} is a complex number manifesting both the stress at the bottom and the phase lag between the stress and orbital velocity. The power γ is estimated in Voronovich et al. (2008) as $\gamma \approx 0.48$. At first sight, it seems easier to determine the values of the parameters δ_{dis}, δ_{fric}, and γ from experimental data rather than from theoretical developments.

Segur et al. (2005) reported that the plane wave solution becomes linearly and nonlinearly stable when small dissipation δ_{dis} is taken into account. The term of linear dissipation in Eq. (4.50) may be illuminated after the following change

$$A(X,T) = Q(X,T)\exp(-\delta_{dis}T). \quad (4.51)$$

Then Eq. (4.50) becomes

$$-i\left(\frac{\partial Q}{\partial T} + C_{gr}\frac{\partial Q}{\partial X}\right) + \beta\frac{\partial^2 Q}{\partial X^2} + \alpha e^{-2\delta_{dis}T}|Q|^2 Q = 0, \quad (4.52)$$

where we set $\delta_{fric} = 0$ to restrict our interest to effects of linear dissipation only. The exponent in (4.52) reduces the nonlinear effect. It is obvious that if the timescale of the dissipation is much larger than other timescales of the problem, then formulas (4.38), (4.39), (4.40), (4.41), and (4.42) are asymptotically valid. The resulting growth rate of perturbations is

$$\sigma_{BF} = -\delta_{dis} + |\Delta K|\sqrt{2\alpha\beta e^{-2\delta_{dis}T}A_0^2 - \beta^2\Delta K^2}. \quad (4.53)$$

The study of Segur et al. (2005) confirms that the radical expression in Eq. (4.53) defines the onset of the modulational instability. The instability is always cancelled when the time interval becomes sufficiently long. The energy transfer from the carrier wave to the sidebands is still possible and may be substantial if

$$\alpha |A_0|^2 \gg \delta_{dis}. \tag{4.54}$$

Nevertheless, the spectral satellites grow for a limited time, and its increase is halted due to the dissipation.

Considering the case of nonlinear wave damping due to bottom friction only ($\delta_{dis} = 0$), the amplitude of the carrier wave decays in a power-law way in contrast to Eq. (4.51). The exponential growth rate has a rather complicated form, but the instability condition is defined by the following expression (one may compare this expression with the radical in Eq. (4.53))

$$\beta \Delta K^2 \left(2\alpha A_0^2 + F_1 \right) - \beta^2 \Delta K^4 - F_2 > 0, \tag{4.55}$$

where α and β are given by Eq. (4.44) for the general three-dimensional case; F_1 and F_2 relate to the action of the bottom stress and are functions of the complex parameter δ_{fric}, and wave amplitude A_0. The dissipation hampers the development of instability and shrinks the corresponding instability domain. The longitudinal perturbations turn out to be the most susceptible to be influenced by bottom friction. Voronovich et al. (2008) estimate that the longitudinal perturbations become stable when the nonlinear term in Eq. (4.50) becomes less than the frictional one

$$\alpha |A_0|^{2-\gamma} <\approx |\delta_{fric}|. \tag{4.56}$$

Since the velocity components decay exponentially at large depths (see formulas (2.58), (2.59), and (2.60)), the bottom friction produces a significant contribution only when the dimensionless depth $K_0 D$ is not large, and becomes unimportant in the deep-water case. For intermediate depths ($K_0 D \sim 1.5$), realistic estimations foresee that the modulational growth may be seriously suppressed by the nonlinear bottom friction or even cancelled at all.

4.2 Rogue Wave Phenomenon within the Framework of the NLS Equation

In what follows, it is convenient to use the dimensionless form of the NLS equation

$$iq_t + q_{xx} + 2|q|^2 q = 0, \tag{4.57}$$

which results from Eq. (4.35) under the following transformations

$$t = \frac{1}{2}\Omega_0 T, \quad x = 2K_0 (X - C_{gr} T), \quad q = \frac{1}{\sqrt{2}} K_0 A^*. \tag{4.58}$$

Eq. (4.57) corresponds to the deep-water case; it is often called the focusing NLS equation with inherent property that the signs between the nonlinear and the dispersive terms are same. Condition (4.38) is satisfied by Eq. (4.57), and hence, modulational instability is possible in this system.

4.2.1 General Solution of the Cauchy Problem

Equation (4.57) is known to be integrable as it was demonstrated by Zakharov and Shabat (1972) with the help of the Inverse Scattering Transform (IST) (see Novikov et al. 1984, Drazin and Johnson 1989). This technique has attributes of the classic Fourier method (spectrum and eigenfunctions) and allows the determination of some explicit exact solutions and an analytical description of model cases. Nevertheless, from the viewpoint of computations it is trickier than the Fourier transform. Two formulations of the IST exist, suggested by Zakharov and Shabat (1972) and Ablowitz et al. (1974), respectively. We will hereafter follow the latter, usually referred to as the Ablowitz-Kaup-Newell-Segur (AKNS) scheme.

Following the AKNS approach, the initial value problem associated with the focusing NLS equation (4.57) is written as follows:

$$\begin{cases} \dfrac{\partial \Psi}{\partial x} = \begin{pmatrix} \lambda & q \\ -q^* & -\lambda \end{pmatrix} \Psi, \\ \dfrac{\partial \Psi}{\partial t} = \begin{pmatrix} a & a_{12} \\ a_{21} & -a \end{pmatrix} \Psi, \end{cases} \qquad (4.59)$$

where

$$\Psi = \begin{pmatrix} \Psi_1 \\ \Psi_2 \end{pmatrix}, \quad \begin{cases} a = i|q|^2 + 2i\lambda^2 \\ a_{12} = iq_x + 2i\lambda q \\ a_{21} = iq_x^* - 2i\lambda q^* \end{cases}.$$

The eigenvalues λ are independent of time and constitute the spectrum. The first matrix equation in (4.59) defines the spatial dependence of the eigenfunctions $\Psi(x,t)$, while the second one defines their time dependence. The solution of the initial value problem consists of determining the spectrum for the initial perturbation $q(x, t = 0)$ (the direct scattering transform), and then restoring the wave field on the basis of the permanent spectrum and known time-dependent eigenfunctions (the IST).

The spatially localized eigenfunctions correspond to the discrete spectrum, while the others form the continuous spectrum. The discrete spectrum is responsible for the existence of solitary waves discovered first for the Korteweg-de Vries equation by Zabusky and Kruskal (1965) and later found in many important equations and observed in different physical problems. The solitons are localized nonlinear solutions that interact elastically with other solitons and quasilinear waves, preserving their energy and shape. Considering the Cauchy problem on the infinite interval with $q \to 0$ when $x \to \pm\infty$, any initial perturbation evolves into a set of solitons (they correspond to the discrete spectrum) and a spreading due to dispersive oscillatory tail (described by the continuous spectrum). Since the system is conservative, the spreading waves decay in amplitude over time, so the solitons represent the asymptotic solution of the initial value problem for the integrable equation such as the NLS equation.

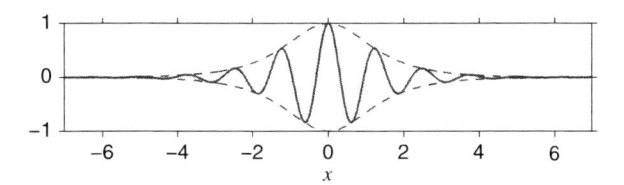

Fig. 4.2 The envelope soliton solution (4.60) with $A_{es} = 1$ and $V_{es} = 10$. The real part of the solution is given by the *solid line*, while the *dashed lines* show $\pm|q_{es}|$

A solitary wave of the NLS equation is represented by the nonlinear envelope as follows

$$q_{es}(x,t) = A_{es} \frac{\exp\left[i\left(xV_{es}/2 - \left((V_{es}/2)^2 - A_{es}^2\right)t\right)\right]}{\cosh(A_{es}(x - V_{es}t))}, \quad (4.60)$$

where A_{es} is the amplitude, and V_{es} is the speed of the envelope soliton. The envelope soliton (4.60) is plotted in Fig. 4.2.

The part of the wave field corresponding to the continuous spectrum tends to the following solution when $t \to \infty$

$$q_{tail}(x,t) = \frac{Q}{\sqrt{t}} \exp\left[i\left(\frac{x^2}{4t} + 2Q^2 \ln t + \Theta\right)\right], \quad (4.61)$$

where Q and Θ are functions of the ratio x/t (Ablowitz and Segur 1979).

The multisoliton solution may be found analytically, but even the two-soliton expression (a bi-soliton) has a rather complicated form (see Peregrine 1983, Akhmediev and Ankiewicz 1997). That is why the numerical solution of the NLS equation (4.57) is often used as the less laborious way of analysis. The nonlinear combinations of solitons (4.60) with background waves will be discussed in Sect. 4.2.3.

4.2.2 Nonlinear-Dispersive Formation of a Rogue Wave

The problem, which is at the heart of our attention in this section, is "How can normal waves evolve into a rogue wave?" Let us draw the reader's attention to the fact that the change $q \to q^*$ in the NLS equation (4.57) is equivalent to the time inversion: $t \to -t$. This property becomes understood from relation (4.20), where the complex conjugation corresponds to inversion of the velocity, which should result in time inversion. Due to this symmetry, instead of considering the process of freak wave generation, the opposite evolution may be investigated. Suppose we know the rogue wave profile. What are the waves resulting from its disintegration? Hence, the problem of seeking the wave combinations causing rogue waves is transformed into an initial value problem for a probable rogue wave shape.

We choose the expected freak wave $q(x)$ having a pulse-like shape. It should be understood, however, that the NLS equation is valid for weakly modulated wave trains, and thus an impulse field $q(x)$ corresponds to a wave group $\eta(X)$ with carrier wavenumber K_0 on the sea surface.

The initial value problem for the NLS equation (4.57) for some shapes of the impulse disturbances was studied by Satsuma and Yajima (1974), Burzlaff (1988), Kaup and Malomed (1995), Desaix et al. (1996), Clarke et al. (2000), and Slunyaev (2001). They provide qualitatively and quantitatively similar results. A particular shape of the expected freak wave

$$q_{fr} = A_p \operatorname{sech}\left(\frac{x}{L}\right), \qquad (4.62)$$

where A_p and L are real positive values, will be considered. The initial value problem (4.59) for the potential (4.62) was solved by Satsuma and Yajima (1974), and the discrete eigenvalues are defined by the expression

$$\lambda_n L = \frac{M}{\pi} - n + \frac{1}{2}, \quad n = 1, 2, \ldots, N_s, \qquad (4.63)$$

where the number of discrete eigenvalues is given by

$$N_s = \left[\frac{M}{\pi} + \frac{1}{2}\right]. \qquad (4.64)$$

The bracket $[f]$ in (4.64) denotes the integer part of f. The parameter M is the "mass" of the initial wave shape

$$M = \int_{-\infty}^{\infty} |q_{fr}| dx, \qquad (4.65)$$

which is equal to $M = \pi A_p L$ for a freak wave having the shape of the *sech* function (4.62). Discrete eigenvalues emerge only when the mass exceeds the threshold value $M \geq M_{th}$ where

$$M_{th} = \frac{\pi}{2}. \qquad (4.66)$$

Every eigenvalue λ corresponds to an envelope soliton (4.60) with parameters defined by the relation

$$\lambda = \frac{1}{2}A_{es} + i\frac{1}{4}V_{es}. \qquad (4.67)$$

Therefore, the integer number N_s is often called the soliton number.

Actually, besides the *sech*-like initial pulse, the solution (4.63) and (4.64) is valid for a variety of real shapes $q_{fr}(x)$ (see Satsuma and Yajima 1974, Burzlaff 1988, Kaup and Malomed 1995, Desaix et al. 1996, Clarke et al. 2000, and Slunyaev 2001). The number M is a convenient parameter of the Cauchy problem since it corresponds to the ratio of nonlinearity with respect to dispersion in Eq. (4.57). This ratio is

$$\frac{q|q|^2}{q_{xx}} \propto |q|^2 L^2 \propto M^2 \qquad (4.68)$$

(L is the characteristic length scale) and shows the significance of nonlinear effects compared with dispersive effects. Note that one envelope soliton has "mass"

4.2 Rogue Wave Phenomenon within the Framework of the NLS Equation

$$M_{es} = \pi, \tag{4.69}$$

that is twice the threshold value in Eq. (4.66). When the number of discrete eigenvalues N_s is large (and M is large, too) then formula (4.64) with M defined by Eq. (4.65) agrees with the quantization rule of Bohr and Sommerfeld (Landau and Lifshitz 1980) for the scattering problem defined by the first equation in (4.59).

Now, two states of the wave evolution may be compared: the expected freak wave and the result of its evolution over time. When solitons emerge, their amplitudes satisfy Eq. (4.67). The maximum amplitude of the solitary part of the field is equal to $A_{max} = 2\lambda_1$. Applying the formal definition of a rogue wave (I.1), as $A_p/A_{max} \geq 2$, one may easily obtain the necessary condition for the freak wave occurrence from (4.63)

$$M \leq \frac{2\pi}{3} \approx 2.1. \tag{4.70}$$

Condition (4.70) allows the existence of no more than one envelope soliton in the wave field (see (4.64)), which may give birth to a freak wave. If $M < M_{th}$, the wave field does not contain solitons at all. Thus, solitons are not necessary for the formation of a freak wave; and what is more important, rogue waves in the form of very nonlinear (with large M) pulse-like wave packets cannot be formed. An intensive dispersive tail is most important in this process. Its asymptotic form is given by (4.61).

4.2.2.1 Case of a Small Mass Parameter

In the limit $M \to 0$, solitons do not appear (actually, when $M < \pi/2, N_s = 0$). Hence, only spreading decaying wave trains may occur as the result of the Cauchy problem. The problem may be considered in the linear approach as a first approximation. The linear wave grouping due to dispersion has been considered in Chap. 3. The evolution of the Gaussian pulse in the linear limit is described by the exact solution (3.27).[1] In the deep-water case, the dispersion law results in quadratic wavenumber modulation, optimal for the dispersive focusing. Other shapes of the expected rogue waves correspond to other distributions of the energy and phases in the dispersive train, although the quadratic phase modulation remains optimal and becomes apparent over time. Note that solution (4.61) has quadratic phase modulation if Q and Θ are taken as constant. In fact, these functions correct the optimal phase modulation, but such a correction becomes less important if t is large. Formula (3.23), describing the asymptotic behavior of the wave field stemming from the linear disintegration of a rogue wave in the form of the delta-function, does not contain these corrections.

In a more complicated case, the expected freak wave profile may be represented by the Gaussian shape with quadratic phase modulation, specified by the parameter β,

$$q_{fr} = A_p \exp\left[-(x/L)^2\right] \exp\left(-i\beta x^2\right). \tag{4.71}$$

---

[1] Note that here there is a temporal wave evolution, while in Chap. 3 it is the spatial one.

Kaup and Malomed (1995) showed that the squared modulation leads to a growth of the thresholds of the emerging soliton (4.66); the discrete eigenvalues move along the real axis closer to zero. This result agrees with simple estimations made in (Slunyaev et al. 2002). Thus, a freak wave, expected as a modulated impulse (4.71), is the result of an even lower number of solitons (i.e., one or none).

4.2.2.2 Competition of the Self-Modulation and Dispersive Effects

When a wave train has both amplitude and phase modulation, the effects of dispersive and nonlinear self-focusing will compete with each other. We illustrate this case with the help of a numerical simulation of the NLS equation (4.57). The initial condition is taken in the following form

$$q(x,t=0) = A_0 \left(1 + \varepsilon \cos\left(x/L_{BF}\right)\right) \exp\left(ix^2/L_{disp}^2\right), \quad (4.72)$$

where $A_0 = 0.043$, $\varepsilon = 0.1$ (it is a small parameter specifying the strength of the amplitude modulation), $L_{BF} = 28$, and L_{disp} varies. Length scales L_{BF} and L_{disp} are responsible for the amplitude and wavenumber modulation, respectively. Results of the numerical simulation of the evolution of the envelope are presented in Fig. 4.3

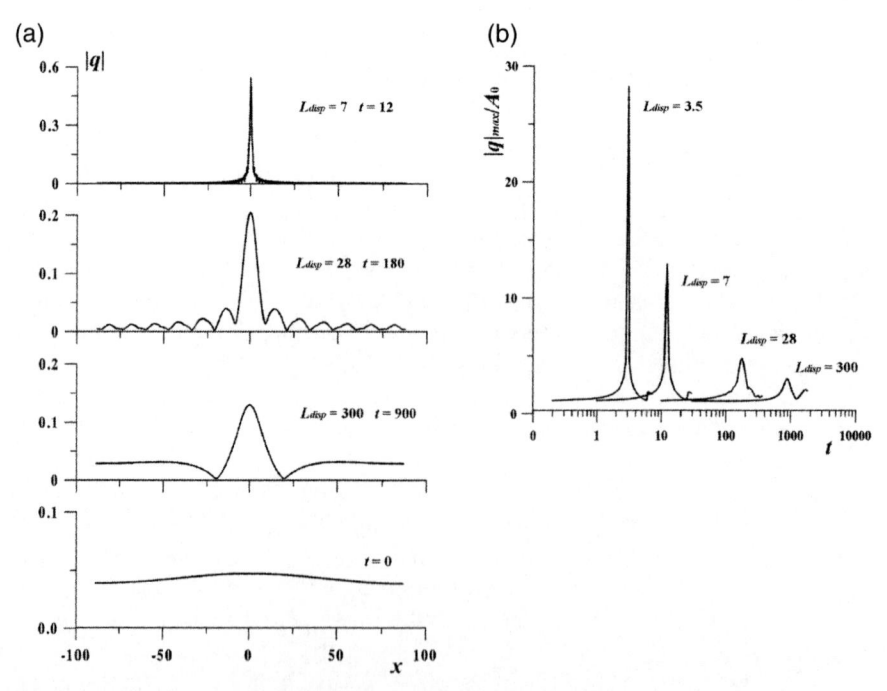

Fig. 4.3 Numerical simulations of the wave train with amplitude and phase modulation (4.72) within the NLS model. (**a**) The initial profile and the amplified wave envelopes for different values of the parameter of phase modulation L_{disp}. (**b**) The maxima of the wave field envelope versus time, corresponding to the cases shown in panel (**a**)

for different values of L_{disp}. Focusing due to the phase modulation (small L_{disp}) happens rapidly and for short time scales, while modulational growth due to the Benjamin-Feir instability (large L_{disp}) occurs for longer time scales. When the modulational instability starts, the growth is exponential and then saturates (see Kharif et al. 2001 and Slunyaev et al. 2002). The dispersive focusing exhibits a power-law dependence and has a sharp maximum (see Sect. 3.2).

4.2.3 Solitons on a Background and Unstable Modes

In the previous section, the Cauchy problem on infinite intervals with zero conditions at infinity has been considered, and the dispersive quasi-linear waves could spread and decay. The case of non-zero background waves, as well as the periodic problem, leads to the nonlinear interaction of the waves of the discrete spectrum with quasi-linear waves that cannot be neglected.

4.2.3.1 Exact Solutions

The so-called "breather" solutions[2] of the NLS equation represent nonlinear interactions of an envelope soliton with a background plane wave. These basic solutions were first obtained by Kuznetsov (1977), Kawata and Inoue (1978), and Ma (1979), and completed later in Peregrine (1983), Akhmediev et al. (1985, 1987), Nakamura and Hirota (1985), Tajiri and Watanabe (1998), Dysthe and Trulsen (1999), Calini and Schober (2002), and Slunyaev et al. (2002). The simplest case of a breather is represented by a single eigenvalue of the modified associated scattering problem when the solution tends to a plane wave at infinity ($x \to \pm\infty$). Except the different boundary conditions, other details of the approach are similar to the classical one given by (4.59). From this point of view, a breather may be called a soliton (usually called Ma soliton) or the superposition of a classical envelope soliton of the NLS equation with a plane wave. Naturalness and richness of this interpretation will be demonstrated below. We will use the general form of this solution, obtained directly from the inverse scattering problem in Slunyaev et al. (2002) and through the Hirota method in Tajiri and Watanabe (1998). For the dimensionless NLS equation (4.57), the solution with a single eigenvalue λ is given by

$$q_{br}(x,t) = e^{2it} \times$$
$$\times \frac{\cos\mu\cos(2\gamma(x-vt)+2i\psi) - \cosh\psi\cosh(2\Gamma(x-V_{br}t)+2i\mu)}{\cos\mu\cos(2\gamma(x-vt)) - \cosh\psi\cosh(2\Gamma(x-V_{br}t))}, \quad (4.73a)$$

---

[2] Note that this name for solutions of this kind is not generally accepted. For instance, Akhmediev and Ankiewicz (1997) refer to the specific collision of two solitons with equal speeds localized at the same place as a "breather."

where

$$\Gamma = -\sinh\psi\cos\mu, \quad \gamma = \cosh\psi\sin\mu,$$
$$V_{br} = -\frac{\cosh 2\psi \sin 2\mu}{\sinh\psi\cos\mu}, \quad v = \frac{\sinh 2\psi \cos 2\mu}{\cosh\psi\sin\mu}, \quad (4.73b)$$
$$\lambda = \cos(\mu + i\psi).$$

Here, V_{br} is the speed of the plane wave perturbation that is traveling as a group. The speed and parameters v, γ, and Γ are defined through the eigenvalue. Solution (4.73) is scaled with respect to the amplitude of the surrounding plane wave ($q_{br}(x,t) \to \exp(2it)$ when $x \to \pm\infty$).

While evolving, the perturbations of the plane wave oscillate with the period

$$T_{br} = \frac{\pi}{\cos 2\mu \sinh 2\psi} \quad (4.74)$$

and stay within the interval

$$|A_{br} - A_{pw}| \le |q_{br}| \le |A_{br} + A_{pw}|, \quad (4.75)$$

where

$$A_{br} = 2\cosh\psi\cos\mu, \quad A_{pw} = 1 \quad (4.76)$$

(A_{pw} denotes the amplitude of the plane wave). The following relations between the breather's and eigenvalue properties may be straightforwardly found from (4.73):

$$\lambda = \frac{A_{br}}{2} - i\sin\mu\sinh\psi \quad \text{and} \quad V_{br} = 4\,\text{Im}\,(\lambda)\frac{1+\coth^2\psi}{2}. \quad (4.77)$$

V_{br} is the breather velocity defined in (4.73b), and A_{br} plays the role of the breather amplitude.

Solution (4.73) may look differently, like a pulsating disturbance (Fig. 4.4a) or like a propagating group of the plane wave perturbations (Fig. 4.4b). It is straightforward to see that in the case $\lambda \in \mathfrak{R}$ the solution (4.73) tends to the time-periodic Ma soliton when $\lambda > 1$,

$$q_{br}(x,t) = e^{2it+i\varphi_0} \times \frac{\cos(\omega_{br}t - 2i\psi) - \cosh\psi\cosh(2\Gamma(x-x_0))}{\cos(\omega_{br}t) - \cosh\psi\cosh(2\Gamma(x-x_0))}, \quad (4.78)$$

where

$$\Gamma = -\sinh\psi, \quad \omega_{br} = \frac{2\pi}{T_{br}}, \quad \lambda = \cosh\psi.$$

Solution (4.78) does not propagate, since $V_{br} = 0$. When $0 < \lambda < 1$, the solution (4.73) is reduced to the Akhmediev et al. (1985) solution[3]

---

[3] Akhmediev et al. (1987) also found a double periodic (in time and space) solution.

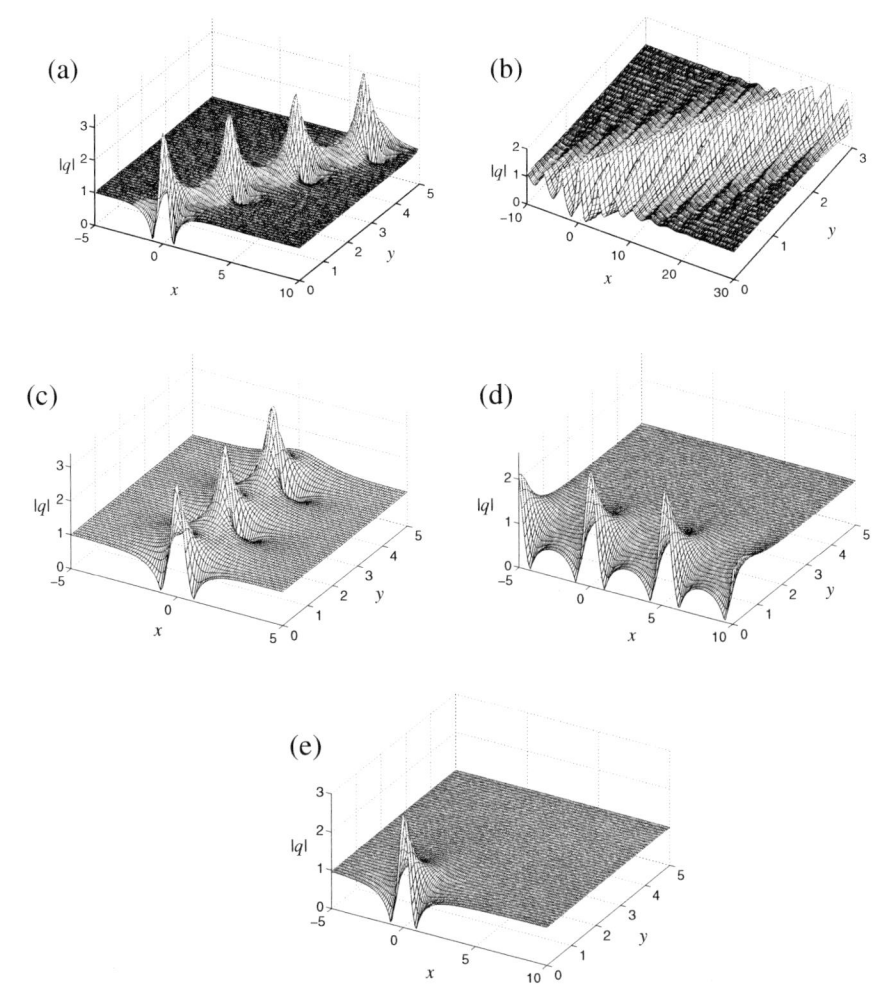

4.2 Rogue Wave Phenomenon within the Framework of the NLS Equation

Fig. 4.4 Breather solutions (4.73) of the NLS equation. (**a**) A traveling Ma 'soliton-like solution ($\lambda = 1.2 + 0.2i$). (**b**) A traveling envelope-like solution ($\lambda = 0.5 + 0.2i$). (**c**) The time-periodic Ma soliton ($\lambda = 1.2$). (**d**) The space-periodic Akhmediev solution ($\lambda = 0.8$). (**e**) The rational solution of Peregrine ($\lambda = 1$)

$$q_{br}(x,t) = e^{2it+i\varphi_0} \times \frac{\cos\mu\cos(2\gamma(x-x_0)) - \cosh(\sigma t - 2i\mu)}{\cos\mu\cos(2\gamma(x-x_0)) - \cosh(\sigma t)}, \quad (4.79)$$

where

$$\gamma = \sin\mu, \quad \sigma = 2\sin(2\mu), \quad \lambda = \cos\mu.$$

The solution (4.79) does not propagate; it is space-periodic and breathes once. The so-called Peregrine (1983) solution is the limit of Eq. (4.73) when $\lambda \to 1$ is imposed:

$$q_{br}(x,t) = e^{2it+i\varphi_0}\left(1 - \frac{4(1+4it)}{1+4(x-x_0)^2+16t^2}\right). \qquad (4.80)$$

Nakamura and Hirota (1985) called this rational solution the explode-decay solitary wave. The examples of the particular solutions (4.78), (4.79), and (4.80) are shown in Fig. 4.4c–e.

Peregrine (1983) pointed out that the Kuznetsov-Ma soliton tends to a usual envelope soliton solution of the NLS equation when its amplitude is much larger than the plane wave amplitude ($A_{br} \gg A_{pw}$). According to formula (4.75), the behavior of the general breathing wave (4.73) may evidently be interpreted in some sense as a linear superposition of a nonlinear envelope with its own amplitude A_{br} and a plane wave with amplitude A_{pw}.

Let us now suppose that the soliton has run away from the region of interaction with the plane wave and is propagating over the zero background (see illustration in Fig. 4.5). Since the nonlinear spectrum λ is conserved, the breather's eigenvalue will be related to the envelope soliton parameters by Eq. (4.67). Comparing Eq. (4.77) with Eq. (4.67), one may conclude how the collision with a plane wave affects a soliton: the envelope preserves its amplitude in the interaction, $A_{br} = A_{es}$, but it accelerates (compare the speeds defined by Eqs. (4.67) and (4.77) for the same value of λ). Figure 4.6 illustrates how combinations of envelope and plane wave parameters result in different kinds of breathing waves. Horizontal curves on the

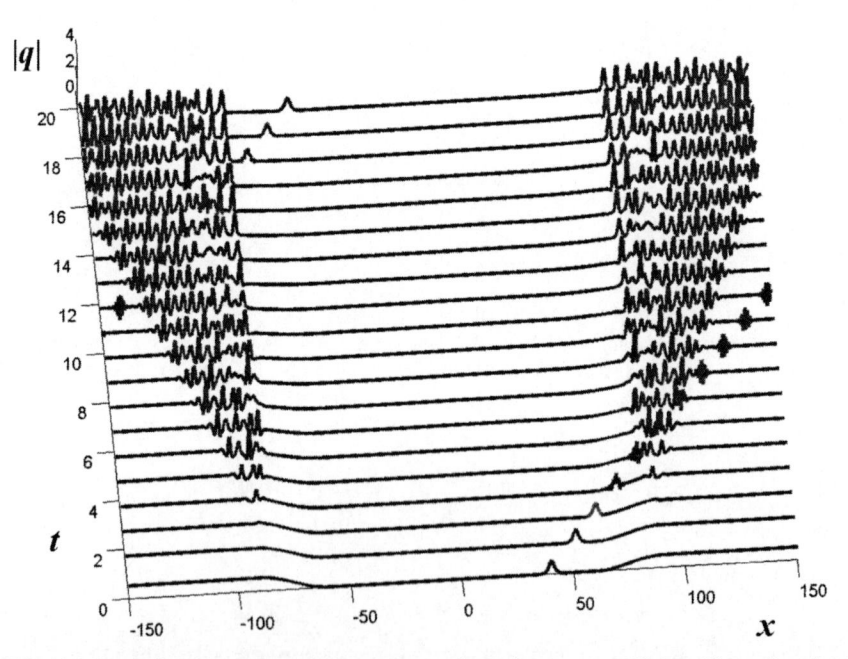

Fig. 4.5 Numerical simulation of an envelope soliton-plane wave collision. The soliton is originally located at the zero background and has rightward velocity, while the plane wave solution ($|x| > 75$) does not move in the chosen frame of reference. Periodic boundary conditions are employed. It is readily seen how the soliton climbs up the plane wave and restores its original shape

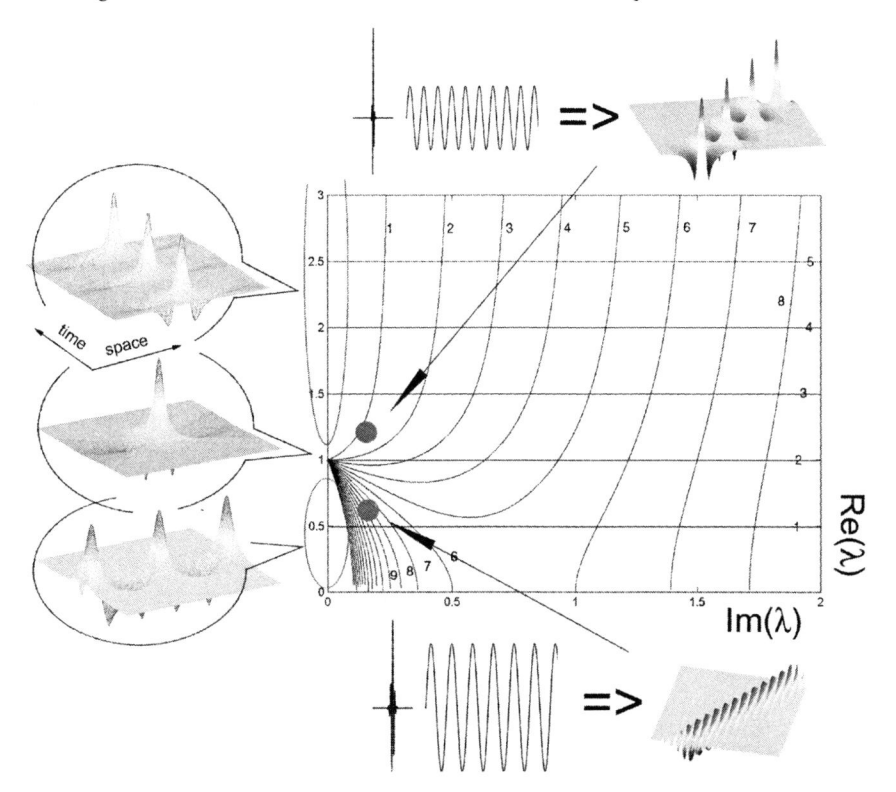

4.2 Rogue Wave Phenomenon within the Framework of the NLS Equation

Fig. 4.6 The λ-plane of an envelope soliton over background and corresponding solutions. Left column of images, from *top* to *bottom*: the time-periodic Ma solution ($\lambda > 1$), the limiting Peregrine solution ($\lambda = 1$), and the space-periodic Akhmediev solution ($\lambda < 1$). Horizontal lines on the plane denote dimensionless soliton amplitudes, and bent curves show the isovelocity lines

λ-plane show the isoamplitude lines (the amplitude values are given by numbers); bent curves represent the isovelocity lines (numbers indicate corresponding values of V_{br}). The traveling solution is less influenced by the plane wave when A_{es} is large and/or the difference between the speeds of the soliton and the plane wave is large (Slunyaev 2006).

It follows from formula (4.75) that when an envelope soliton interacts with the background plane wave, the maximum wave field is just the linear superposition of the amplitudes of the soliton and the background wave. The maximum wave amplification that can be achieved in this process (3 times) is obtained with the Peregrine soliton (4.80) (Fig. 4.4e) that presents a single oscillation of one localized perturbation of the plane wave. This solution corresponds to the case $A_{es}/A_{pw} = 2$.

4.2.3.2 Chaotic Behavior of the Wave Modulations

When several breathing waves interact (they may be called multibreathers) more complex solutions have been considered (see Calini and Schober 2002). Besides the complexity of the analytical description (and comprehension of the dynamics by eye), the case of interacting breathers ("solitonic turbulence;" see Zakharov et al. 2006a,b) is sensitive with respect to any kind of perturbations. This is due to the fact that the NLS breathers are homoclinic orbits of the equation—thus, small perturbations (for example, such as the round-off errors of numerical computations) may result in chaotic behavior of the wave modulations (see Ablowitz and Herbst 1990, Ablowitz et al. 2000, 2001). The integrable NLS equation possesses the Fermi-Pasta-Ulam recurrence (Newell 1981), although its approximate models may lose this property. Therefore, the detailed description of real modulations of sea waves obviously fails, if the cases of many breathers or evolution over time are considered. The statistical approach given in Sect. 4.4 may turn out to be more successful, although the soliton and breather conceptions are often very useful for the understanding of particular wave dynamics.

It is straightforward to show that the Akhmediev solution (4.79) provides an exponential growth, which is equal to the modulational growth rate (4.40) (here values $\alpha = 2, \beta = 1, A_0 = 1, \Delta K = 2\gamma$ should be employed). Hence, the breather solutions indeed describe the development of the Benjamin-Feir instability. The wavenumber corresponding to the maximum growth rate (4.41) results in the length scale $\gamma = 1/\sqrt{2}$, hence, $\mu = \pi/4$. This corresponds to the breather amplitude (4.76) $A_{br} = \sqrt{2}$. The amplification factor achieved by solution (4.79) is then obtained using (4.75) and reads

$$\frac{\max(|q_{br}|)}{A_{pw}} = \frac{A_{br} + A_{pw}}{A_{pw}} = 1 + \sqrt{2} \approx 2.4, \qquad (4.81)$$

which is smaller than the result of the Peregrine solution (3 times) but still agrees with the rogue wave criterion (I.1).

In such a way the breathing solutions are closely linked with the modulational instability. They are often associated with *unstable modes* that can be revealed in the wave field with the help of the IST and then used to describe the modulational properties of the waves. Osborne et al. (2005) suggested this approach on the basis of the scattering problem on a periodic domain (long before, the IST was applied by Osborne and Petti (1994) to analyze shallow water laboratory waves). The unstable modes were also studied by Islas and Schober (2005); another way to use the IST to analyze real sea waves was suggested and developed in papers by Slunyaev et al. (2005, 2006) and will be considered in Sect. 4.7.1.

When the statistical description is concerned, the effects of nonlinear instabilities do influence the probability distribution functions. These effects are beyond the bound nonlinear wave corrections and certainly result from the dynamics described in this section. Some recent results on sea wave probabilistic descriptions that involve nonlinear wave-wave interactions and the bridge to the dynamical aspect will be discussed further in Sect. 4.4.

To conclude this section, we briefly present some results on chaos and modulational instabilities that go beyond the NLS equation. Solving the Zakharov equation numerically, Caponi et al. (1982) discovered that owing to the modulational instability the Stokes wave train evolves into a chaotic system. They called this phenomenon *confined chaos*. Later, Yasuda and Mori (1997) simulated numerically the long-term evolution of a perturbed Stokes wave by modulational instabilities using the fully nonlinear equations. They showed that the evolution of the perturbed Stokes wave trains into Fermi-Pasta-Ulam recurrence or chaos depends on the number of Fourier modes within the unstable range, the initial steepness of the Stokes waves and the nonlinear coupling between the fundamental modes and higher harmonics of the modulation. By means of high-order modeling with sufficiently many degrees of freedom, they demonstrated that Stokes wave trains evolve into chaotic systems. The numerical method used by Yasuda and Mori is due to Dommermuth and Yue (1987) and is presented in the next section.

4.3 Rogue Wave Simulations within the Framework of the Fully Nonlinear Equations

In the previous section, the dynamics of rogue waves have been investigated within the framework of weakly nonlinear theories. The validity of these models can become questionable in accurately describing rogue waves that are strongly nonlinear water waves. The approximate models may be inaccurate when the extreme wave event is occurring. Hence, to have a more realistic description of this phenomenon, it is necessary to use the fully nonlinear equations (2.13), (2.28), (2.30), and (2.31) with initial and boundary values for the potential and elevation. In constant depth and infinite depth, the bottom condition is given by Eqs. (2.46) and (2.61), respectively. Most of the time, these equations are solved numerically. Different numerical methods are available for the spatio-temporal evolution of water-wave groups. Among the many papers devoted to extreme wave events due to modulational instability and dispersive or directional focusing, one can cite the following list, which is not exhaustive: Henderson et al. (1999), Bateman et al. (2001), Clamond and Grue (2002), Touboul et al. (2006), Clamond et al. (2006), Fochesato et al. (2007), Dyachenko and Zakharov (2005), and Kharif et al. (2008). Among the different kinds of numerical methods used commonly to simulate unsteady evolution of strongly nonlinear free surface flows due to modulational instability and dispersive or directional focusing, we present here a High-Order Spectral Method (HOSM) and a Boundary Integral Equation Method (BIEM).

4.3.1 A High-Order Spectral Method

We consider the case of infinite depth and introduce the following dimensionless variables into Eqs. (2.13), (2.28), (2.29), and (2.61): $x = K_0 X, y = K_0 Y, z = K_0 Z, \zeta = K_0 \eta, \varphi = \phi \cdot (g/K_0^3)^{-1/2}, p = P/(\rho g/K_0)$, where K_0 is a reference wavenumber. Hence, the kinematic and dynamic boundary conditions become

$$\frac{\partial \zeta}{\partial t} + \frac{\partial \varphi}{\partial x}\frac{\partial \zeta}{\partial x} + \frac{\partial \varphi}{\partial y}\frac{\partial \zeta}{\partial y} - \frac{\partial \varphi}{\partial z} = 0 \quad \text{on} \quad z = \zeta, \tag{4.82}$$

$$\frac{\partial \varphi}{\partial t} + \frac{1}{2}\nabla\varphi \cdot \nabla\varphi + p_a + z = 0 \quad \text{on} \quad z = \zeta. \tag{4.83}$$

Following Zakharov (1968), we introduce the velocity potential at the free surface $\varphi^s(x,y,z,t) = \varphi(x,y,z=\zeta(x,y,t),t)$ into Eqs. (4.82) and (4.83)

$$\frac{\partial \zeta}{\partial t} = -\nabla\varphi^s \cdot \nabla\zeta + w\left[1 + (\nabla\zeta)^2\right], \tag{4.84}$$

$$\frac{\partial \varphi^s}{\partial t} = -\zeta - \frac{1}{2}\nabla\varphi^s \cdot \nabla\varphi^s + \frac{1}{2}w^2\left[1 + (\nabla\zeta)^2\right] - p_a, \tag{4.85}$$

with

$$w = \frac{\partial \varphi}{\partial z}(x,y,z=\zeta(x,y,t),t). \tag{4.86}$$

The main difficulty is the computation of the vertical velocity at the free surface, w. Following Dommermuth and Yue (1987), the potential $\varphi(x,y,z,t)$ is written as a finite perturbation series up to a given order M

$$\varphi(x,y,z,t) = \sum_{m=1}^{M} \varphi^{(m)}(x,y,z,t). \tag{4.87}$$

The term $\varphi^{(m)}$ is of $O(\varepsilon^m)$ where ε, a small parameter, is a measure of the wave steepness. Then, expanding each $\varphi^{(m)}$ evaluated at $z = \zeta$ in a Taylor series about $z = 0$, we obtain

$$\varphi^s(x,y,t) = \sum_{m=1}^{M} \sum_{l=0}^{M-m} \frac{\zeta^l}{l!} \frac{\partial^l}{\partial z^l} \varphi^{(m)}(x,y,z=0,t). \tag{4.88}$$

At a given instant of time, φ^s and ζ are known so that from Eq. (4.88), we can calculate $\varphi^{(m)}$ at each order:

$$O(1): \varphi^{(1)}(x,y,z=0,t) = \varphi^s(x,y,t), \tag{4.89}$$

$$O(m): \varphi^{(m)}(x,y,z=0,t) = -\sum_{l=1}^{m-1} \frac{\zeta^l}{l!}\frac{\partial^l}{\partial z^l}\varphi^{(m-l)}(x,y,z=0,t). \quad m \geq 2. \tag{4.90}$$

These boundary conditions, with the Laplace equations $\Delta\varphi^{(m)}(x,y,z,t) = 0$ to be solved in the domain occupied by the water, define a series of Dirichlet problems for $\varphi^{(m)}$.

For 2π-periodic conditions in (x,y) in deep water, $\varphi^{(m)}$ can be written as follows

$$\varphi^{(m)}(x,y,z,t) = \sum_{j=0}^{\infty}\sum_{l=0}^{\infty} \varphi_{jl}^{(m)}(t)\exp(k_{jl}z)\exp[i(jx+ly)], \tag{4.91}$$

where $k_{jl} = \sqrt{j^2 + l^2}$.

4.3 Rogue Wave Simulations within the Framework of the Fully Nonlinear Equations

Note that $\varphi^{(m)}(x,y,z,t)$ automatically satisfies the Laplace equation and the condition $\lim \nabla \varphi^{(m)}(x,y,z,t) \to 0$ as $z \to -\infty$.

For constant finite depth d, an alternative decomposition must be used, namely

$$\varphi^{(m)}(x,y,z,t) = \sum_{j=0}^{\infty} \sum_{l=0}^{\infty} \varphi_{jl}^{(m)}(t) \frac{\cosh\left[k_{jl}(z+d)\right]}{\cosh(k_{jl}d)} \exp\left[i(jx+ly)\right]. \quad (4.92)$$

Substitution of (4.91) into the set of Eqs. (4.89) and (4.90) gives the modes $\varphi_{jl}^{(m)}(t)$. The vertical velocity at the free surface is then

$$w = \sum_{m=1}^{M} \sum_{l=0}^{M-m} \frac{\zeta^l}{l!} \frac{\partial^{l+1}}{\partial z^{l+1}} \varphi^{(m)}(x,y,z=0,t). \quad (4.93)$$

Substitution of Eq. (4.93) into the boundary conditions (4.84) and (4.85) yields the evolution equations for φ^s and ζ.

The numerical method used to solve the evolution equations (4.84) and (4.85) is similar to that developed by Dommermuth and Yue (1987). Equations (4.84) and (4.85) are integrated using a pseudo-spectral treatment with $N = JL$ wave modes, where $J = max(j)$ and $L = max(l)$ and retaining nonlinear terms up to order M. Once the surface elevation $\zeta(x,y,t)$ and the potential at the free surface $\varphi^s(x,y,z,t)$ at time t are known, the modal amplitudes may be computed. The spatial derivatives of $\varphi^{(m)}$, φ^s, ζ, and w are calculated in the spectral space, while nonlinear terms are evaluated in the physical space at a discrete set of collocation points (x_j, y_l). Fast Fourier Transforms (FFTs) are used to link spectral and physical spaces. Equations. (4.89) and (4.90) are solved in the spectral space. Evolution equations for φ^s and ζ are integrated in time using a fourth-order Runge-Kutta integrator with constant time step. The calculation accuracy depends on several sources of errors due to truncation in the number of modes J and L, and order M, aliasing phenomenon, numerical time integration, etc. Numerical convergence tests can be found in Dommermuth and Yue (1987) and Skandrani et al. (1996).

Another version of HOSM developed by West et al. (1987) can also be used. The difference between both methods lies in the way we compute w from $\varphi^{(m)}$. West et al. (1987) assume a power series for w as

$$w(x,y,t) = \sum_{m=1}^{M} w^{(m)}, \quad (4.94)$$

where

$$w^{(m)} = \sum_{l=0}^{m-1} \frac{\zeta^l}{l!} \frac{\partial^{l+1}}{\partial z^{l+1}} \varphi^{(m-l)}(x,y,z=0,t). \quad (4.95)$$

The treatment of nonlinear terms in the latter method is useful for comparisons between the truncated fully nonlinear equation and approximate models, such as the Zakharov equation.

4.3.2 A Boundary Integral Equation Method

In this section, we describe a 2D numerical wave tank based on a boundary integral equation method applied to rogue waves due to energy focusing in a small area. The computational domain is defined as a volume of fluid bounded by a bottom, two lateral walls, a paddle, and the free surface (Fig. 4.7). The boundary corresponding to the bottom, lateral walls, and paddle is denoted by $\partial \Omega_{SB}$ while the free surface is denoted by $\partial \Omega_{FS}$. The Laplace equation (2.13) is solved within this domain. The no-flux condition along the solid boundaries $\partial \Omega_{SB}$ is

$$\frac{\partial \varphi}{\partial n} = \mathbf{v}_{SB} \cdot \mathbf{n}, \tag{4.96}$$

where $\mathbf{v}_{SB}$ is the velocity of the rigid boundaries set equal to zero on the bottom and vertical walls and equal to the velocity of the paddle located at the beginning of the numerical wave tank. The unit normal to the boundaries is $\mathbf{n}$.

On the free surface, $\partial \Omega_{FS}$, the potential $\varphi(x,z,t)$ satisfies the kinematic boundary condition written in the following form

$$\frac{D\mathbf{r}}{Dt} = \nabla \varphi, \tag{4.97}$$

with $\mathbf{r} = (x,z)^t$. The dynamic boundary condition (2.29) is rewritten as

$$\frac{D\varphi}{Dt} = \frac{1}{2}\nabla \varphi \cdot \nabla \varphi - gz - p_a. \tag{4.98}$$

Hence the set of equations to be solved is the Laplace equation $\Delta \varphi = 0$ in the fluid domain, Eq. (4.96) on the rigid boundaries and Eqs. (4.97) and (4.98) on the free surface. These equations are solved numerically using a boundary integral equation method (BIEM) and a mixed Euler-Lagrange (MEL) time marching scheme. Green's second identity transforms the Laplace equation into the following boundary integral equation for φ

$$\int_{\partial \Omega} \varphi(P) \frac{\partial G}{\partial n}(P,Q) d\partial \Omega - \int_{\partial \Omega} \frac{\partial \varphi}{\partial n}(P) G(P,Q) d\partial \Omega = \alpha(Q) \varphi(Q), \tag{4.99}$$

Fig. 4.7 Sketch of the computational domain for the numerical wave tank

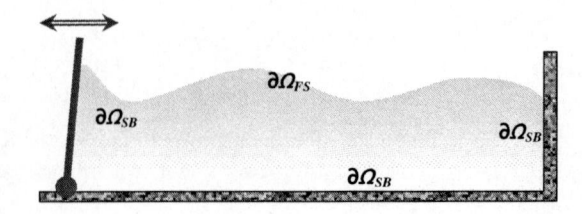

4.3 Rogue Wave Simulations within the Framework of the Fully Nonlinear Equations

where the integration includes both the solid and free surfaces $\partial\Omega = \partial\Omega_{SB} \cup \partial\Omega_{FS}$, G is the 2D free-space Green's function, P and Q denote two points of the fluid domain, and **n** is the outward unit vector normal to the boundary. The angle $\alpha(Q)$ is defined as follows: $\alpha(Q) = 0$ or -2π when Q is outside or inside the fluid domain, respectively, and $\alpha(Q) = \theta$ when Q is on the boundary. The angle θ is the inner angle with respect to the fluid domain at point Q along the boundary.

Eq. (4.99) can be written in a more explicit form.

For the free surface $Q \in \partial\Omega_{FS}$:

$$\theta\varphi - \int_{\partial\Omega_{FS}} \varphi(P) \frac{\partial G}{\partial n}(P,Q) d\partial\Omega + \int_{\partial\Omega_{SB}} \frac{\partial\varphi}{\partial n}(P) G(P,Q) d\partial\Omega$$
$$= \int_{\partial\Omega_{SB}} \varphi(P) \frac{\partial G}{\partial n}(P,Q) d\partial\Omega - \int_{\partial\Omega_{FS}} \frac{\partial\varphi}{\partial n}(P) G(P,Q) d\partial\Omega, \quad (4.100)$$

For the solid boundaries $Q \in \partial\Omega_{SB}$:

$$\int_{\partial\Omega_{FS}} \varphi(P) \frac{\partial G}{\partial n}(P,Q) d\partial\Omega + \int_{\partial\Omega_{SB}} \frac{\partial\varphi}{\partial n}(P) G(P,Q) d\partial\Omega$$
$$= \theta\varphi + \int_{\partial\Omega_{SB}} \varphi(P) \frac{\partial G}{\partial n}(P,Q) d\partial\Omega - \int_{\partial\Omega_{FS}} \frac{\partial\varphi}{\partial n}(P) G(P,Q) d\partial\Omega. \quad (4.101)$$

The unknowns are $\partial\varphi/\partial n$ on $\partial\Omega_{FS}$ and φ on $\partial\Omega_{SB}$. The above equations, that are assumed to be satisfied at a discrete set of points on the boundary of the fluid domain, are transformed into a linear system of algebraic equations for a finite number of unknowns (for more details see the Appendix). Equations (4.97) and (4.98) are integrated in time using a fourth-order Runge-Kutta integrator.

4.3.3 Numerical Simulation of Rogue Waves Due to Modulational Instability

Henderson et al. (1999) investigated the time evolution of a 2D almost uniform wave train with a small growing modulation. They performed numerical experiments—it was observed that energy focuses into a short group of steep waves, called steep wave events (SWE). Details about the numerical code used to study water wave modulations can be found in Dold (1992). It was found that the breather solutions of the NLS equation fit numerical SWEs rather well. These SWEs are considered to be rogue waves. Hence, the rogue-wave mechanism due to the Benjamin-Feir instability is confirmed in fully nonlinear computations. Later, Clamond and Grue (2002) and Clamond et al. (2006) performed fully nonlinear numerical simulations of lengthy evolution of a 2D localized long-wave packet. The numerical method used is a fast converging iterative solution of the Laplace equation. One part of the solution is obtained by FFT, while another part is highly nonlinear and consists of integrals with kernels that decay quickly in space (see Clamond and Grue 2001). The result showed how interacting solitary wave groups that emerge from the long wave packet can produce rogue wave events (see Sect. 4.3.5). Dyachenko and Zakharov (2005) and Zakharov et al. (2006b) claimed that rogue wave events are due to solitonic turbulence emerging from modulational instability of Stokes waves. This scenario seems similar to that suggested by Clamond and Grue (2002) and Clamond et al. (2006). Their simulation was based on a numerical method using a conformal mapping of the fluid domain to the lower half plane. More generally, this quasi-solitonic turbulence can appear as a result of the instability of narrow spectral distributions of gravity waves. More details on solitonic turbulence can be found in Zakharov et al. (2006a).

Until now, studies on rogue waves have not taken into account the action of wind. Previous works on rogue waves have not considered the direct effect of wind on their dynamics. It was assumed that they occur independently of wind action, far away from storm areas where wind wave fields are formed. Kharif et al. (2008) considered wind above rogue waves, both numerically and experimentally. Two kinds of mechanisms yielding rogue waves were investigated. In this subsection, we present numerical experiments showing how a rogue event can arise from the modulational instability of a Stokes' wave train with and without wind.

In different situations, several authors have experimentally investigated the influence of wind on the evolution of mechanically generated gravity-water waves. Bliven et al. (1986), Li et al. (1987), and Waseda and Tulin (1999) studied the influence of wind on Benjamin-Feir instability. Contrary to results reported by Bliven et al. (1986) and Li et al. (1987), Waseda and Tulin (1999) found that wind did not suppress the sideband instability. Banner and Song (2002) numerically studied the onset of wave breaking in nonlinear wave groups in the presence of wind forcing. Here, we investigate how wind forcing modifies unforced extreme wave events due to modulational instability.

The generation of extreme wave events can be simply obtained from the Benjamin-Feir instability (or modulational instability) of uniformly traveling trains of Stokes' waves in deep water. It is well-known that these trains are subject to sideband instability producing amplitude and frequency modulations. This instability corresponds to a quartet interaction between the fundamental component (the carrier) $\mathbf{K}_0 = K_0(1, 0)$ counted twice, and two satellites $\mathbf{K}_1 = K_0(1+p, q)$ and $\mathbf{K}_2 = K_0(1-p, -q)$, where pK_0 and qK_0 are the longitudinal wavenumber and transversal wavenumber of the modulation, respectively.

As was emphasized in Sect. 4.1.1, the dominant instability of a uniformly-traveling train of Stokes' waves in deep water is the 2D modulational instability (class I) provided that its steepness is less than $s = 0.30$. For higher values of the wave steepness, 3D instabilities (class II) become dominant, phase locked to the unperturbed wave. First we shall focus on the 2D nonlinear evolution of a Stokes' wave train suffering modulational instability without wind action, and then with wind action. Two series of numerical simulations that can be found in Kharif et al. (2008) are presented. They correspond to two wave trains of five and nine waves, respectively.

4.3.3.1 Rogue Waves without Wind Action

A series of 2D rogue-wave simulations in deep water, obtained when using the numerical method described in Sect. 4.3.1, is presented. The wind effect on the water-wave dynamics is neglected, hence the atmospheric pressure, p_a, is set equal to zero in Eq. (4.83).

First, we consider the case of wave trains of five waves. The initial condition is a Stokes wave train of steepness $s = 0.11$, disturbed by its most unstable perturbation, which corresponds to $p \approx 1/5$. The fundamental wavenumber of the Stokes wave is chosen so that integer numbers of the sideband perturbation (satellites) can be fit into the computational domain. For the considered case, the normalized[4] dimensionless fundamental wave harmonic of the Stokes' wave is $k_0 = 5$ and the dominant side bands are $k_1 = 4$ and $k_2 = 6$ for the subharmonic and superharmonic part of the perturbation, respectively. The wave parameters have been rescaled so that the wavelength of the perturbation is equal to 2π. There exist higher harmonics involved in the interactions, which are not presented here. The normalized amplitude of the perturbation relative to the Stokes wave amplitude is initially taken to be equal to 10^{-3}. The order of nonlinearity in the HOSM is $M = 6$; the number of mesh points is greater than $(M+1)k_{max}$, where k_{max} is the highest harmonic taken into account in the simulation. To compute the evolution length of the wave train, the time step is chosen to be equal to one hundredth of the fundamental period of the basic wave, T_0. In this way, the time step satisfies the Courant-Friedrichs-Levy (CFL) condition.

The time histories of the normalized amplitude of the carrier, lower sideband, and upper sideband of the most unstable perturbation are plotted in Fig. 4.8a. Another perturbation that was initially linearly stable becomes unstable in the vicinity of the maximum of modulation, resulting in the growth of the sidebands $k_3 = 3$ and $k_4 = 7$. The nonlinear evolution of the 2D wave train exhibits the Fermi-Pasta-Ulam recurrence phenomenon. This phenomenon is characterized by a series of modulation-demodulation cycles in which initially uniform wave trains become modulated and then demodulated until they are again uniform. Here, one cycle of modulation-demodulation is reported. At time $t \approx 360T_0$, the initial condition is more or less recovered.

At the maximum of modulation $t = 260T_0$, one can observe a temporary frequency (and wavenumber) downshifting since the subharmonic mode $k_1 = 4$ is dominant. At this stage, a very steep wave occurs in the group as it can be seen in Fig. 4.9a. Notice that the solid line represents the free surface without wind effect while the dotted line corresponds to the case with wind effect, which will be discussed later. Figure 4.9b–d shows the free surface profiles at several instants in time. The solid lines correspond to the case without wind action. We can emphasize that no breaking occurs during the numerical simulation. Dold and Peregrine (1986) have numerically studied the nonlinear evolution of various modulating wave trains towards breaking or recurrence. For a given number of waves in the wave train, breaking always occurs above a critical initial steepness, and below, a recurrence

---

[4] Note the wavenumbers in this Section are normalized in a different way than those in Sect. 4.2.

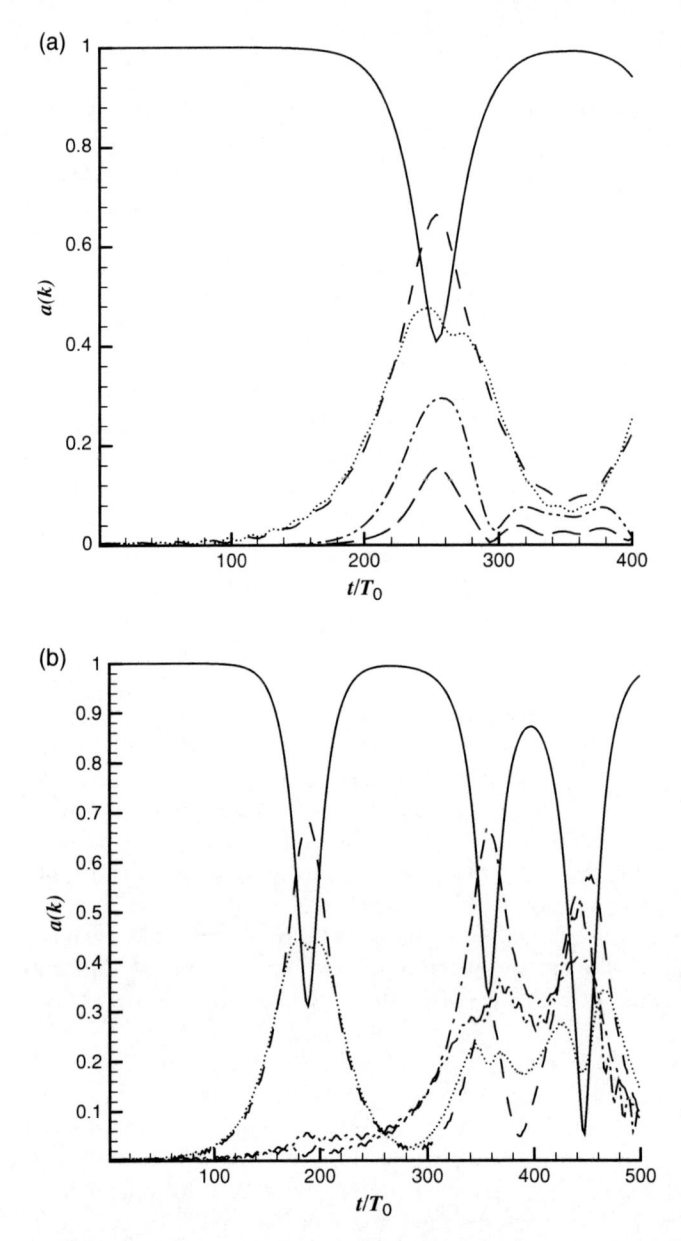

Fig. 4.8 Time histories of the amplitude of the main spectral modes for an evolving perturbed Stokes wave with fundamental wave period T_0, without wind action. (**a**) The fundamental mode $k_0 = 5$ (*solid line*), subharmonic mode $k_1 = 4$ (*dashed line*), superharmonic mode $k_2 = 6$ (*dotted line*). The initial wave steepness is $s = 0.11$. The two lowest curves (*dot-dot-dashed and dot-dashed lines*) correspond to the modes $k_3 = 3$ and $k_4 = 7$. (**b**) The fundamental mode $k_0 = 9$ (*solid line*), subharmonic modes, $k_1 = 7$ (*dashed line*) and $k_3 = 8$ (*dot-dashed line*), and superharmonic modes, $k_2 = 11$ (*dotted line*) and $k_4 = 10$ (*dot-dot-dashed line*). The initial wave steepness is $s = 0.13$

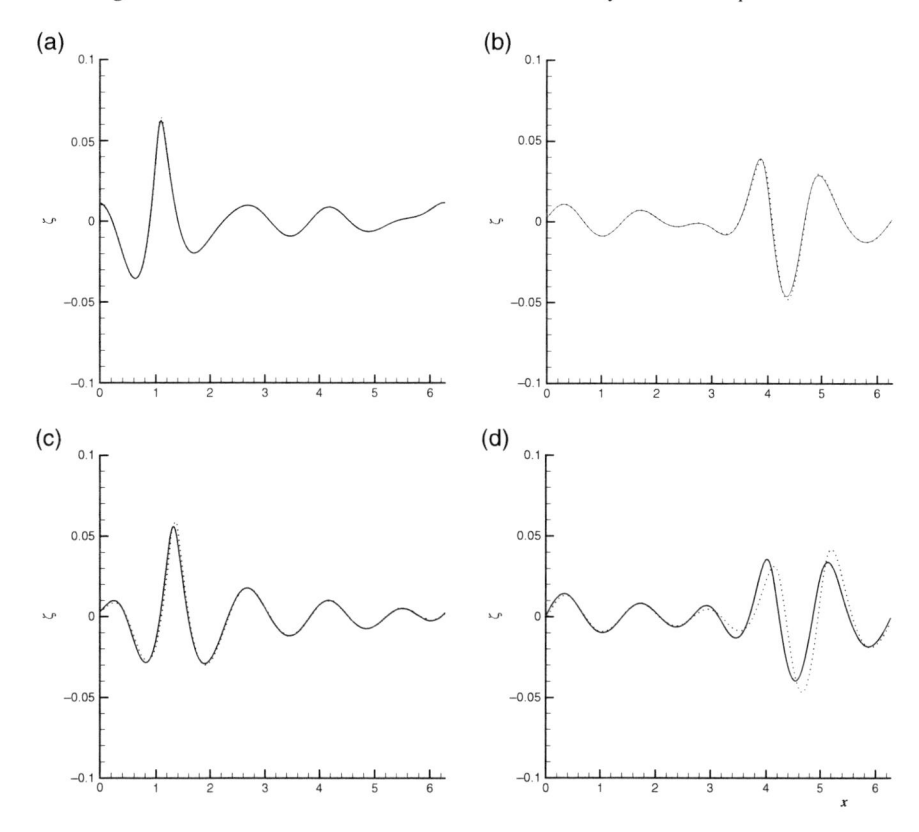

4.3 Rogue Wave Simulations within the Framework of the Fully Nonlinear Equations

Fig. 4.9 Surface wave profiles at $t = 260T_0$ (**a**), $t = 265T_0$ (**b**), $t = 270T_0$ (**c**) and $t = 275T_0$ (**d**): without wind (*solid lines*) and with wind (*dotted lines*)

towards the initial wave group is observed. This problem was revisited by Banner and Tian (1998) who, however, did not consider the excitation at the maximum modulation of the perturbation corresponding to $p \approx 2/5$.

A second numerical simulation corresponding to the case of wave trains of nine waves is now considered. The initial condition is a Stokes wave of steepness $s = 0.13$, disturbed by its most unstable perturbation, which corresponds to $p \approx 2/9$. The unstable sideband perturbation corresponding to $p = 1/9$ is introduced, as well. Hence, we consider the nonlinear evolution of the wave train when two unstable modulations are now present, whereas in the previous case only one unstable modulation was introduced. The fundamental wave harmonic of the Stokes wave is now $k_0 = 9$, and the dominant sidebands are $k_1 = 7$ and $k_2 = 11$ for the subharmonic and superharmonic parts of the perturbation, respectively, while the satellites $k_3 = 8$ and $k_4 = 10$ are the sidebands of the unstable perturbation corresponding to $p = 1/9$. The time histories of the normalized amplitude of the carrier, lower sideband, and upper sideband of the two unstable perturbations are plotted in Fig. 4.8b. A kind of Fermi-Pasta-Ulam recurrence can be observed, which is stopped at $t \approx 500T_0$ by the onset of breaking. Here, the onset of breaking is delayed by the presence of two unstable perturbations. This result is in agreement with those of Dold and Peregrine (1986) and Banner and Tian (1998). At $t = 192\ T_0$, $t = 360\ T_0$, and $t = 445\ T_0$, which correspond to the first, second, and third maxima of modulation, an extreme wave event occurs as shown in Fig. 4.10a (solid line), Fig. 4.10e,f. The subharmonic sideband, $k_1 = 7$, is dominant and a temporary frequency downshifting is observed.

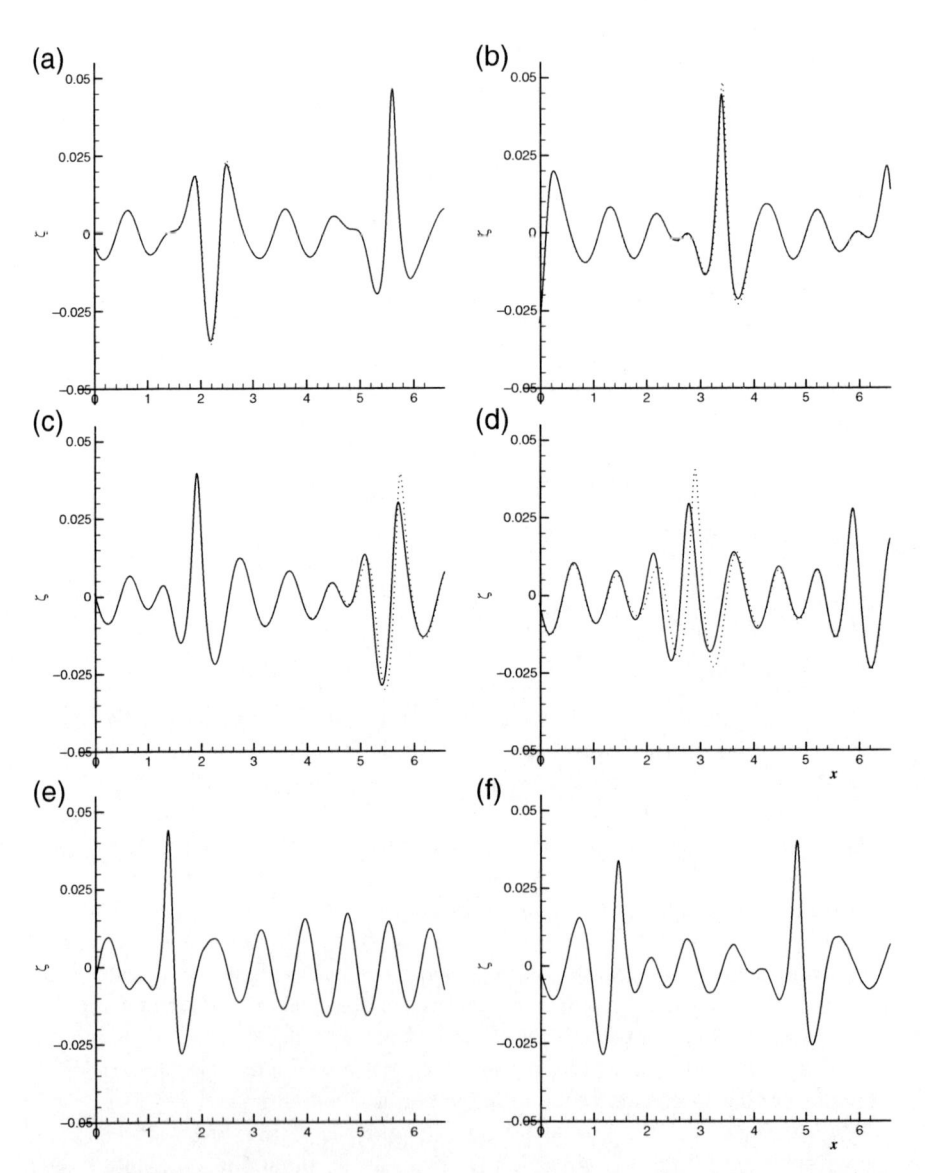

Fig. 4.10 Surface wave profiles at $t = 192\ T_0$ (**a**), $t = 195\ T_0$ (**b**), $t = 200\ T_0$ (**c**), $t = 210\ T_0$ (**d**), $t = 360\ T_0$ (**e**) and $t = 445\ T_0$ (**f**): without wind (*solid lines*) and with wind (*dotted lines*)

4.3 Rogue Wave Simulations within the Framework of the Fully Nonlinear Equations   127

Figure 4.10b–d gives the profiles of the wave train at $t = 195\ T_0$, $t = 200\ T_0$, and $t = 210\ T_0$, respectively.

Owing to a mode competition between the satellites of the two unstable disturbances, it is now the subharmonic sideband, $k_3 = 8$, of the initially less unstable perturbation that is dominant at the second maximum of modulation.

4.3.3.2 Rogue Waves with Wind Action

Here, we investigate how wind forcing modifies unforced extreme wave events due to the modulational instability. The questions are: how do the extreme wave events due to the modulational instability under wind action evolve? How are the amplification and time duration of these waves under wind effect modified?

It was shown experimentally (Kharif et al. 2008) that steep wave events occurring in wave groups are accompanied by air flow separation. The experimental results are presented in Sect. 4.5. Jeffreys (1925) suggested that the energy transfer from wind to water waves was due to the form drag associated with the air flow separation occurring on the leeward side of the crests. The air flow separation produces a pressure asymmetry with respect to the wave crest resulting in a wave growth. However, this mechanism can be invoked only if the waves are sufficiently steep. For weak or moderate steepness of the waves, this phenomenon cannot apply and the Jeffreys sheltering mechanism becomes irrelevant. Hence, a modified sheltering effect has been suggested by Kharif et al. (2008). Following Jeffreys, the relationship between the pressure at the interface and the local wave slope is given by

$$P_a = \rho_a s_j \left(U_w - C_{ph}\right)^2 \frac{\partial \eta}{\partial X}, \qquad (4.102)$$

where s_j is termed the sheltering coefficient, U_w is the wind speed, C_{ph} is the wave phase velocity, and ρ_a is the density of the air. Expression (4.102) is applied for only steep waves—i.e., when the local wave slope $\partial \eta / \partial X$ becomes larger than a given threshold $(\partial \eta / \partial X)_c$, otherwise $P_a = 0$.

Figure 4.11a,b is similar to Fig. 4.8a,b, respectively, except that now water waves evolve under wind action. Wind forcing is applied over crests of the group of five waves of slopes larger than $(\partial \eta / \partial X)_c = 0.405$, while for the group of nine waves it is applied over crests of slopes steeper than 0.5125. These conditions are satisfied for $256\ T_0 < t < 270\ T_0$ for the first wave train, and for $187\ T_0 < t < 200\ T_0$ and $237\ T_0 < t < 240\ T_0$ for the second—that is, during the maximum of modulation that corresponds to the formation of the extreme wave event. When the values of the wind velocity are too high, the numerical simulations fail during the formation of the extreme wave event, owing to breaking. During the breaking wave process, the slope of the surface becomes infinite, leading numerically to a spread of energy into high wavenumbers. This local steepening is characterized by a numerical blow-up (for methods dealing with an Eulerian description of the flow). To avoid a wave breaking too early, the wind velocity U_w is fixed close to $1.75\ C_{ph}$. Owing to the weak effect of the wind on the kinematics of the crests on which it acts, the phase

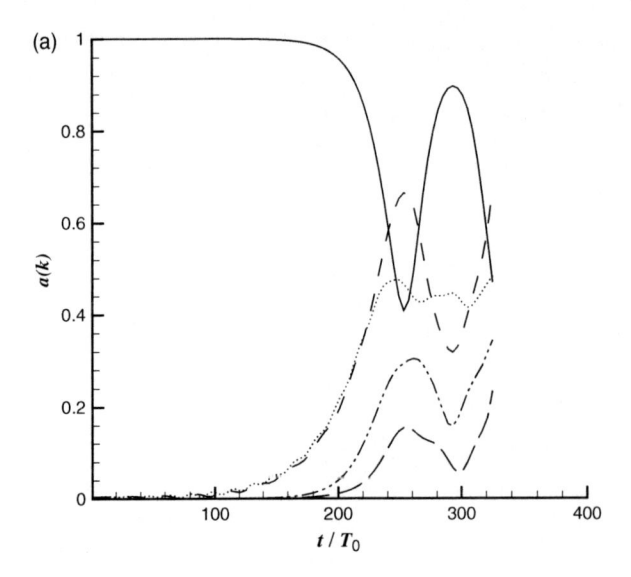

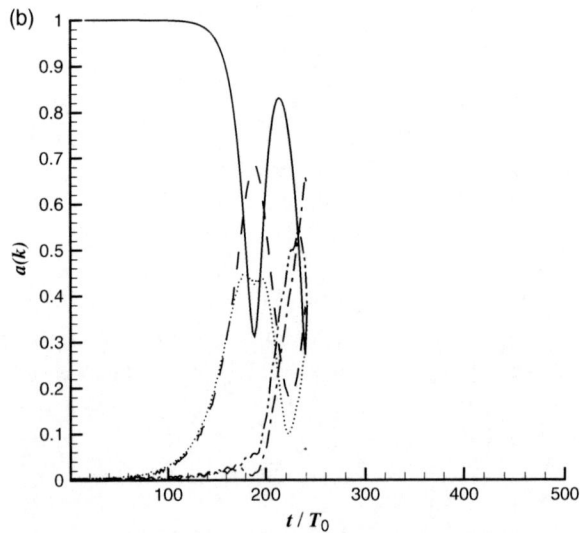

Fig. 4.11 Time histories of the amplitude of the main spectral modes for an evolving perturbed Stokes wave with fundamental wave period T_0, with wind action ($U_w = 1.75 C_{ph}$). (**a**) The fundamental mode $k_0 = 5$ (*solid line*), subharmonic mode $k_1 = 4$ (*dashed line*), and superharmonic mode $k_2 = 6$ (*dotted line*). The initial wave steepness is $s = 0.11$. The two lowest curves (*dot-dot-dashed and dot-dashed lines*) correspond to the modes $k_3 = 3$ and $k_4 = 7$. (**b**) The fundamental mode $k_0 = 9$ (*solid line*), subharmonic modes, $k_1 = 7$ (*dashed line*) and $k_3 = 8$ (*dot-dashed line*), and superharmonic modes, $k_2 = 11$ (*dotted line*) and $k_4 = 10$ (*dot-dot-dashed line*). The initial wave steepness is $s = 0.13$

4.3 Rogue Wave Simulations within the Framework of the Fully Nonlinear Equations 129

velocity, C_{ph}, is computed without wind. The effect of the wind significantly reduces the demodulation cycle and thus sustains extreme wave event.

This feature is clearly shown in Fig. 4.12a,b corresponding to the wave trains of five and nine waves, respectively. The amplification factor is stronger in the presence of wind, and the rogue wave criterion is satisfied during a longer period of time. In the presence of wind forcing, extreme waves evolve into breaking waves at

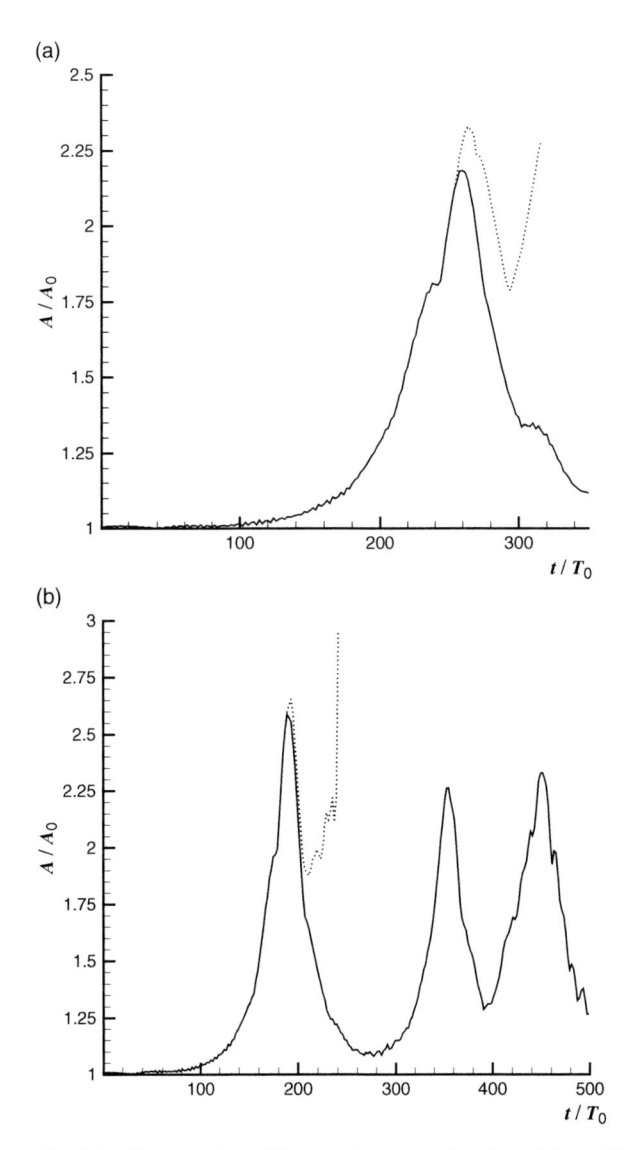

Fig. 4.12 Numerical amplification factor as a function of time without wind (*solid lines*) and with wind (*dotted lines*) for $U_w = 1.75 C_{ph}$: (**a**) for the wave train of five waves, (**b**) for the wave train of nine waves

$t \approx 330\, T_0$ and $t \approx 240\, T_0$ (dotted lines in Fig. 4.12a,b) for wave trains of five and nine waves, respectively. For the case of a wave train of five waves, Fig. 4.9a–d display water wave profiles at different instants of time in the vicinity of the maximum of modulation with and without wind. The solid lines correspond to waves propagating without wind, whereas the dotted lines represent the wave profiles under wind action. These figures show that the wind does not modify the phase velocity of the very steep waves while it increases their height and duration. A similar behavior is shown in Fig. 4.10a–d corresponding to the group of nine waves.

We can conclude that extreme waves occurring under wind action in both wave trains present the same features. Furthermore, in the presence of local wind forcing, extreme waves evolve into breaking waves for initial wave trains of steepness $s = 0.11$ and $s = 0.13$ considered here. In another context, Banner and Song (2002) investigated numerically the onset and strength of breaking for deep water waves under wind forcing and surface shear. In their study, wind modeling is based on Miles' theory, which is different from Jeffreys' sheltering mechanism used in this chapter.

4.3.4 Numerical Simulation of Rogue Waves Due to Dispersive Focusing in the Presence of Wind and Current

As shown in Chap. 3, extreme wave events can be generated by means of dispersive enhancement of wave trains. This mechanism is based upon the dispersive nature of water waves. We consider a chirped wave packet with the leading waves having a higher frequency than trailing waves. For this purpose, the numerical wave tank described in Sect. 4.3.2 is used to produce an extreme wave event.

Within the framework of infinite depth and linear waves, the frequency imposed to the wavemaker located at $X = 0$ is given by formula (3.18), where X_f and T_f are the coordinates of the point of focus in the (X,T) plane. The coordinates of the focus point read

$$T_f = \Delta T \frac{f_{max}}{f_{max} - f_{min}}, \qquad (4.103)$$

$$X_f = \frac{g \Delta T}{4\pi} \frac{1}{f_{max} - f_{min}}, \qquad (4.104)$$

where f_{max} and f_{min} are the maximum and minimum values of the frequency (note that the relation between the cyclic frequency Ω and the frequency f is $\Omega = 2\pi\, f$) imposed to the wavemaker during a period of time ΔT.

Within the framework of the linear theory, the focus points are singular points where the amplitude becomes infinite and behaves as $(T_f - T)^{-1/2}$ (see (3.19)). As it was shown by Touboul et al. (2006) and Kharif et al. (2008), when nonlinear effects are introduced, the rogue wave formation mechanism is not suppressed. In this case, the amplitude of the extreme wave event is finite. The frequency of the wavemaker

4.3 Rogue Wave Simulations within the Framework of the Fully Nonlinear Equations

of the numerical wave tank is varied linearly from $f_{max} = 1.85$ Hz to $f_{min} = 0.8$ Hz during $\Delta T = 23.5$ s. The focusing mechanism is investigated with and without wind as well (Touboul et al. 2006, Kharif et al. 2008). A series of numerical simulations has been run for two values of the wind velocity: $U_w = 0$ m/s and $U_w = 6$ m/s. For each value of the wind velocity, the amplification factor A of the group between fetches X and 1 m is defined as follows

$$A(X, U_w) = \frac{H_{max}(X, U_w)}{H_{ref}}, \quad (4.105)$$

similar to the abnormality index. In (4.105), $H_{max}(X, U_w)$ is the maximum height between two consecutive crests and troughs in the transient group, and the height H_{ref} of the quasi uniform wave train generated at the entrance of the tank is measured at 1 m. Figure 4.13 shows the experimental and computed surface elevation as a function of time at fetch $X = 1$ m. The experimental results will be presented in detail in Sect. 4.5.

Using definition (4.105), Fig. 4.14 describes the spatial evolution of the numerically computed amplification factor. For a value of the threshold wave slope fixed to be equal to 0.3, a blow-up of the numerical simulation occurs owing to the onset of breaking. This threshold value is too low and the transfer of energy from the wind to the steep waves leads to wave breaking. The threshold value of the slope beyond which the wind forcing is applied has been increased and is 0.4. This value corresponds to a wave close to the limiting form for which the modified Jeffreys theory applies. The observed asymmetry between the focusing and defocusing regimes

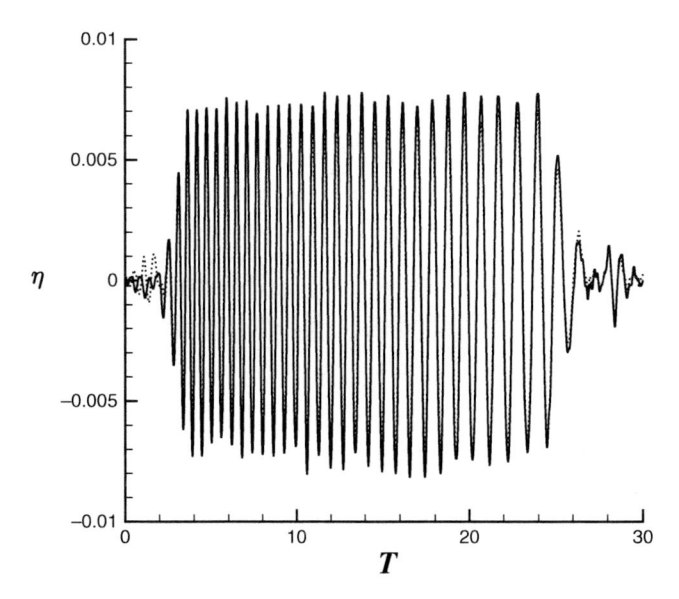

Fig. 4.13 Surface elevation as a function of time at fetch $X = 1$ m: experiments (*solid line*) and numerical simulation (*dotted line*) within the framework of the spatio-temporal focusing

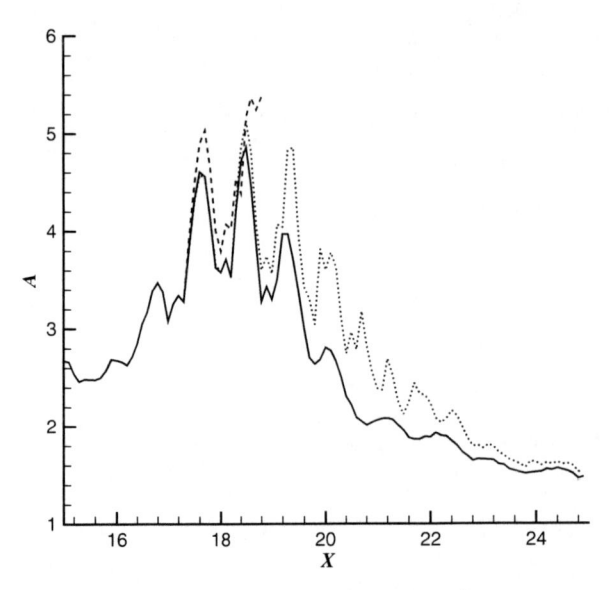

Fig. 4.14 Numerical amplification factor $A(X, U_w)$ as a function of the distance X for two values of the wind velocity within the framework of the spatio-temporal focusing: $U_w = 0$ (*solid line*) and $U_w = 6$ m/s with the threshold value of the wave slope taken to be equal to 0.4 (*dotted line*), $U_w = 6$ m/s with the threshold value of the wave slope taken to be equal to 0.3 (*dashed line*)

can be explained as follows. Without wind, the amplitude of the extreme wave is decreasing during defocusing. In the presence of wind, the modified Jeffreys mechanism that is acting locally in time and space amplifies only the highest waves and hence delays their amplitude decrease during the very beginning of the defocusing stage. The competition between the dispersive nature of the water waves and the local transfer of energy from the wind to the extreme wave event leads to a balance of these effects at the maximum of modulation. This asymmetry results in an increase in the lifetime of the steep wave event, which increases with the wind velocity. Hence, the duration of the wind effect is relatively too short to increase the amplification of the extreme wave event significantly. However, a weak increase of the amplification factor is observed in the presence of wind. The main effect of Jeffreys' sheltering mechanism is to sustain the coherence of the short group involving the steep wave event.

Figure 4.15 shows the numerical amplification factor as a function of the normalized fetch X/X_f, where X_f is the abscissa of the point of focus without wind. The experimental amplification factor is plotted for comparison as well. We can observe an excellent agreement between the numerical and experimental results. The numerical and experimental values of the abscissa of the focus point, X_f, and amplification factor, A, are almost the same.

In the presence of wind of velocity $U_w = 6$ m/s, Fig. 4.16a demonstrates that the numerical and experimental amplification factors deviate from one another beyond the focus point. For a value of the threshold wave slope fixed to be equal to 0.4, the

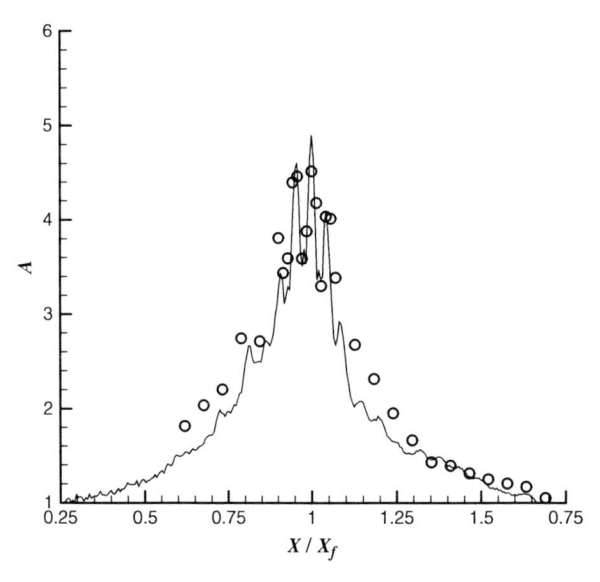

4.3 Rogue Wave Simulations within the Framework of the Fully Nonlinear Equations 133

Fig. 4.15 Numerical (*solid line*) and experimental (*circles*) amplification factor $A(X/X_f, U_w)$ as a function of the normalized distance without wind within the framework of the spatio-temporal focusing

Jeffreys' sheltering mechanism is not effective enough in the present case, whereas a reduction of the threshold value to 0.30 produces the onset of breaking at the focus point.

Wind waves are generally propagating in the presence of a current. Figure 4.16b corresponds to the spatio-temporal focusing in the presence of wind and current with a value of the threshold slope taken to be equal to 0.3. The wind velocity is $U_w = 6$ m/s and a uniform following current corresponding to 2% of U_w has been introduced to have the numerical value of the focus point equal to the experimental value. Generally, the current induced by wind is equal to 3% of the wind velocity. More information about the introduction of a current in the model can be found in the paper by Touboul et al. (2007), who considered the formation of rogue waves from transient wave trains propagating on a current. The laboratory experiments of Wu and Yao (2004) should also be reviewed. The introduction of the following current prevents the onset of breaking. During extreme wave events, the wind-driven current may play a significant role in the wind-wave interaction. The combined action of the Jeffreys sheltering mechanism and wind-driven current may sustain longer extreme wave events. We can see good agreement between the numerical simulation and the experiment. The steep wave event is propagating over a longer distance (or period of time) in the numerical simulation as well as experiments.

To summarize, we can claim that within the framework of the spatio-temporal focusing (or dispersive focusing) both numerical and experimental results are in qualitative good agreement even if some quantitative differences have been observed, namely when the wind-induced current is ignored. Moreover, the importance of a following current on the evolution of the wave group has been emphasized as well.

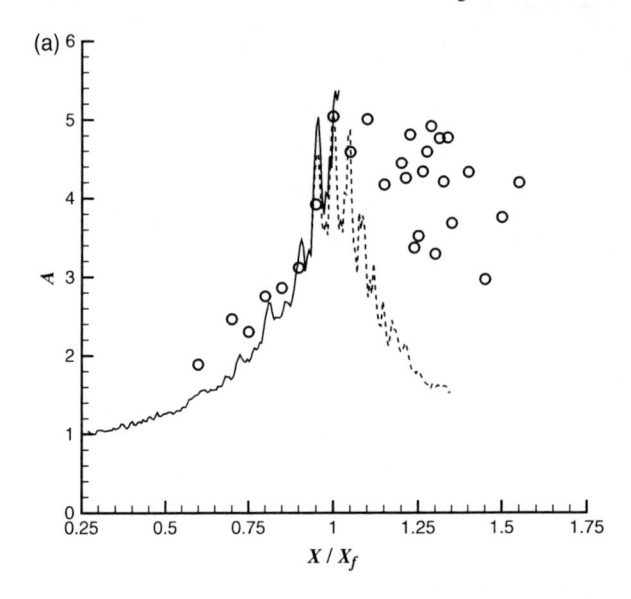

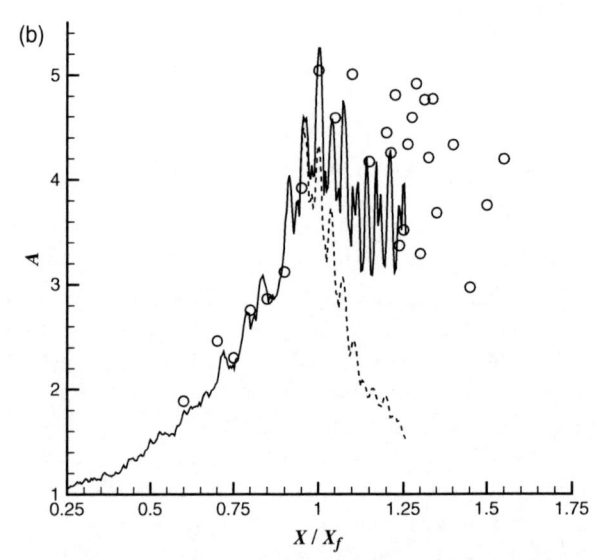

Fig. 4.16 Numerical (*solid and dashed lines*) and experimental (*circles*) amplification factor $A(X/X_f; U_w)$ as a function of the normalized distance for threshold values of the wave slope equal to 0.3 (*solid line*) and 0.4 (*dashed line*) within the framework of the spatio-temporal focusing: (**a**) with wind ($U_w = 6$ m/s), (**b**) in the presence of wind ($U_w = 6$ m/s) and following current

The results of this section have shown that extreme wave events generated by dispersive focusing behave similarly to those due to modulational instability in the presence of wind, as discussed previously. It is found that extreme wave events generated by two different mechanisms exhibit the same behavior in the presence of wind.

4.3.5 Numerical Simulation of Rogue Waves Due to Envelope-Soliton Collision

As it has been discussed in Sect. 4.2, the nonlinear wave groups, also called envelope quasi-solitons, are often a very convenient model for describing the dynamics of modulated waves. The Cauchy problem for an initial localized wave packet was considered by Clamond and Grue (2002) and Clamond et al. (2006). The evolution of a wave packet with an initial steepness of $s = 0.09$ and a bell-shaped (*sech* function) profile has been computed for more than 3,000 wave periods (see Fig. 4.17). During this period of time, three large wave events occur. At about 1,200 wave periods, the wave field consists of three separate solitary wave groups with ordered heights, the steepest being ahead. Until 3,000 wave periods (and later), the groups separate slowly, each group traveling with its characteristic speed. Figure 4.18 illustrates the difference between the weakly nonlinear models (the NLS and extended Dysthe equation) and the fully nonlinear simulation based on Clamond & Grue's scheme regarding envelope dynamics. The analytical theory of the NLS equation predicts that any symmetric envelope (with uniform wavenumber within the group) disintegrates into a finite number of solitons that propagate with the same speed, the linear group velocity, and a small oscillatory tail. For the initial condition considered, it predicts the formation of three solitons that are attached to each other. Furthermore, the corresponding envelope always remains symmetric with respect to the center of the wave group. These bound solitons describe very mild modulations of the envelope amplitude (very long period of recurrence).

The NLS equation predicts the rise of three envelope solitons from the considered initial condition, which is in agreement with the fully nonlinear solution. The shape of each well-separated solitary wave group fits the analytical NLS envelope solutions pretty well. Hence, the observed wave dynamics can be reasonably explained as nonlinear interactions between three perturbed NLS solutions. Nevertheless, it should be noted that the speed of each solitary wave group is not equal to the linear group velocity, as predicted by the NLS theory.

From a qualitative point of view, a somewhat better agreement is obtained with the extended Dysthe equation (Dysthe 1979, Trulsen and Dysthe 1996, Trulsen et al. 2000). This model predicts the early stages of the group splitting (until 300 wave periods) and the characteristic features of the evolution rather well, namely the separation into solitary wave groups and temporary frequency downshifting. However, this model fails to predict the lengthy scenario based on fully nonlinear predictions. Clamond et al. (2006) emphasize that it may be worthwhile to develop a generalization of Dysthe equations, including higher (quintic) nonlinear terms to improve the accuracy and increase the time period of validity.

The result of the fully nonlinear simulation is compared with the fitted exact solution of the NLS equation (the time periodic breather) at the instants of time 155, 156, 157, and 158 wave periods (see Fig. 4.19). This corresponds to the moment of the first steep wave event shown in Figs. 4.17 and 4.18. It is seen that the analytical solution captures some features of the solution rather well and may be used as "first approximation."

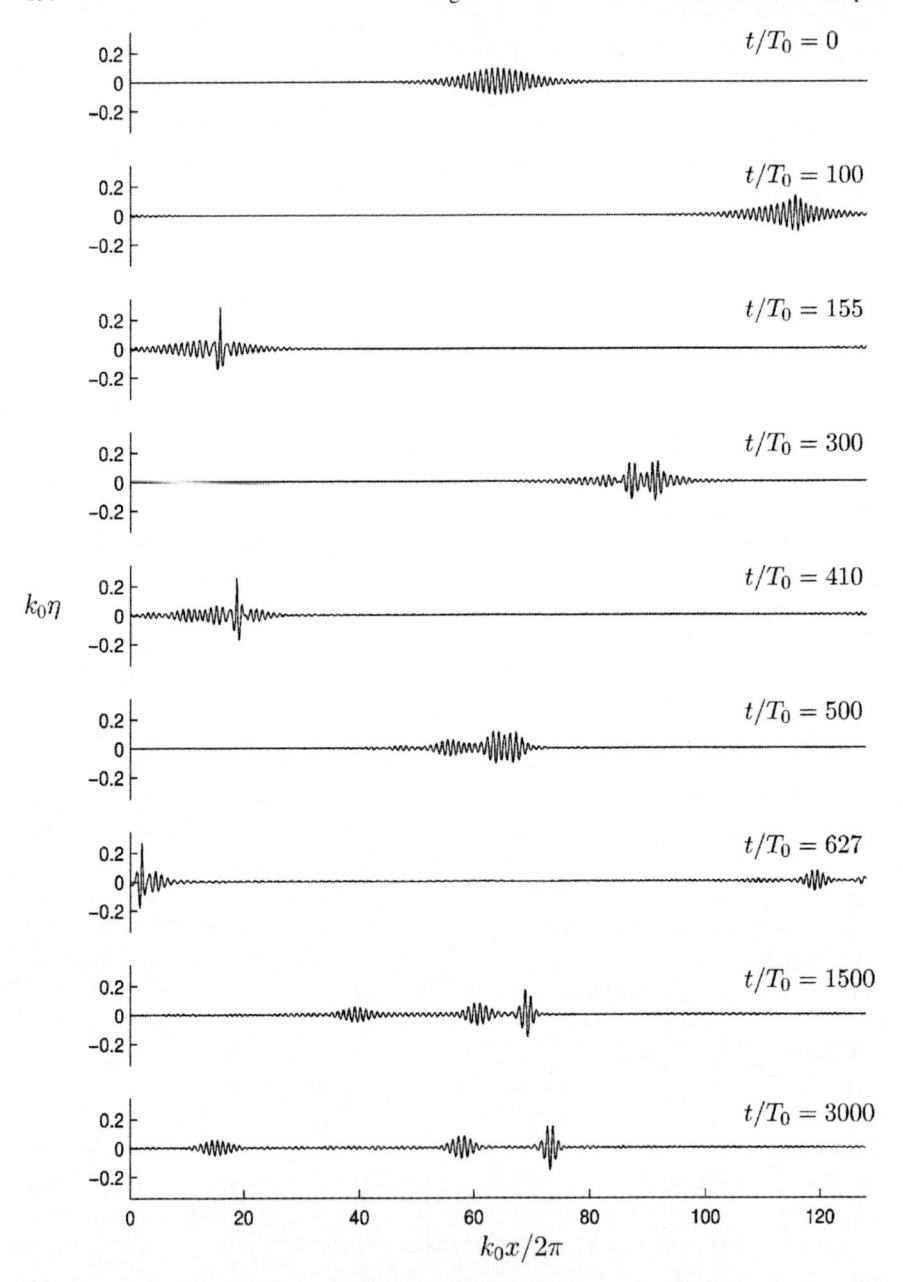

Fig. 4.17 Dynamics of a wave packet with initial *sech*-like shape. k_0 and T_0 denote the carrier wavenumber and period, respectively (see Clamond et al. 2006)

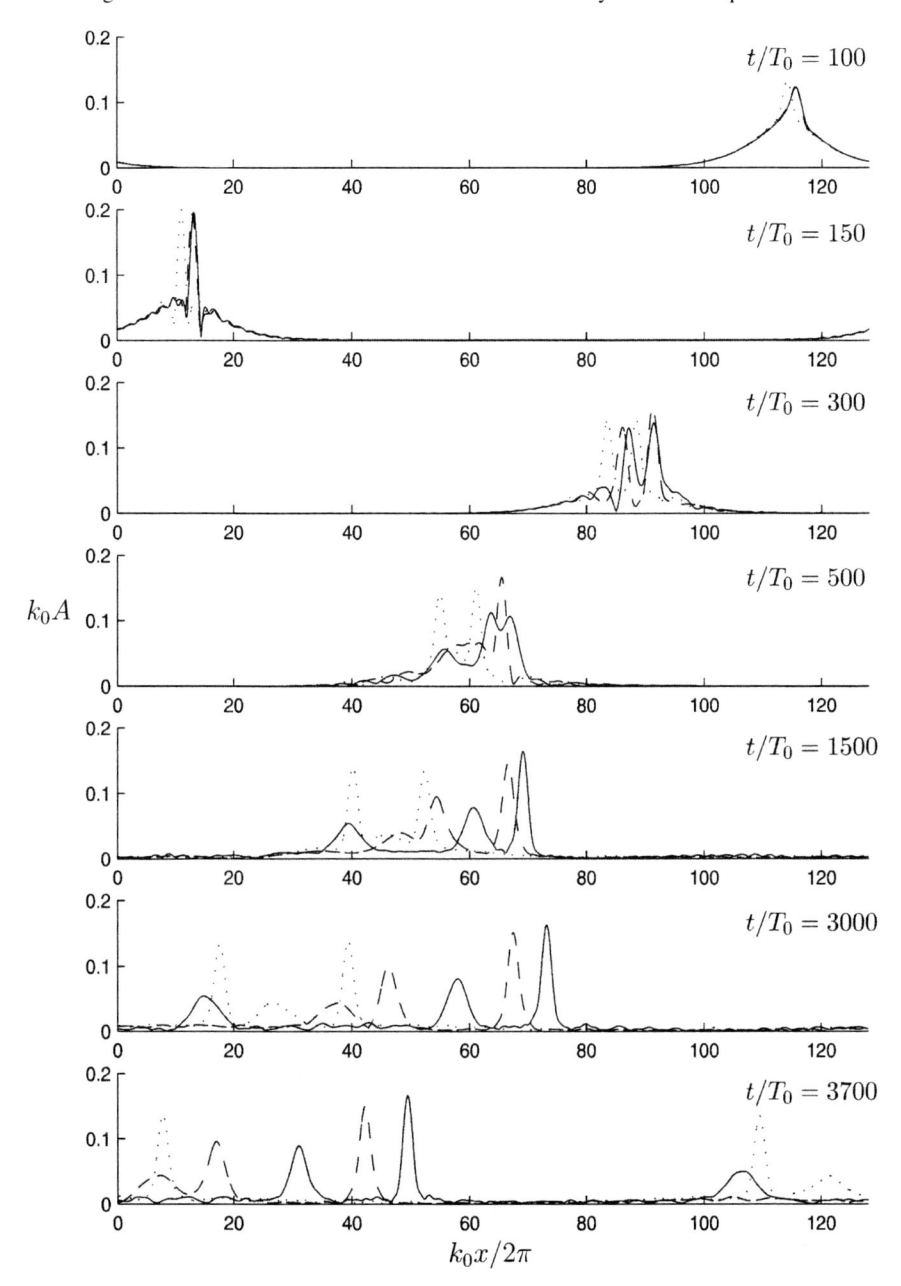

4.3 Rogue Wave Simulations within the Framework of the Fully Nonlinear Equations 137

Fig. 4.18 Comparison of the envelope dynamics from Fig. 4.17 (*solid*) with the results provided by the extended Dysthe equation (*dashed*) and the NLS equation (*dots*), k_0 and T_0 denote the carrier wave number and period, respectively (see details in Clamond et al. 2006)

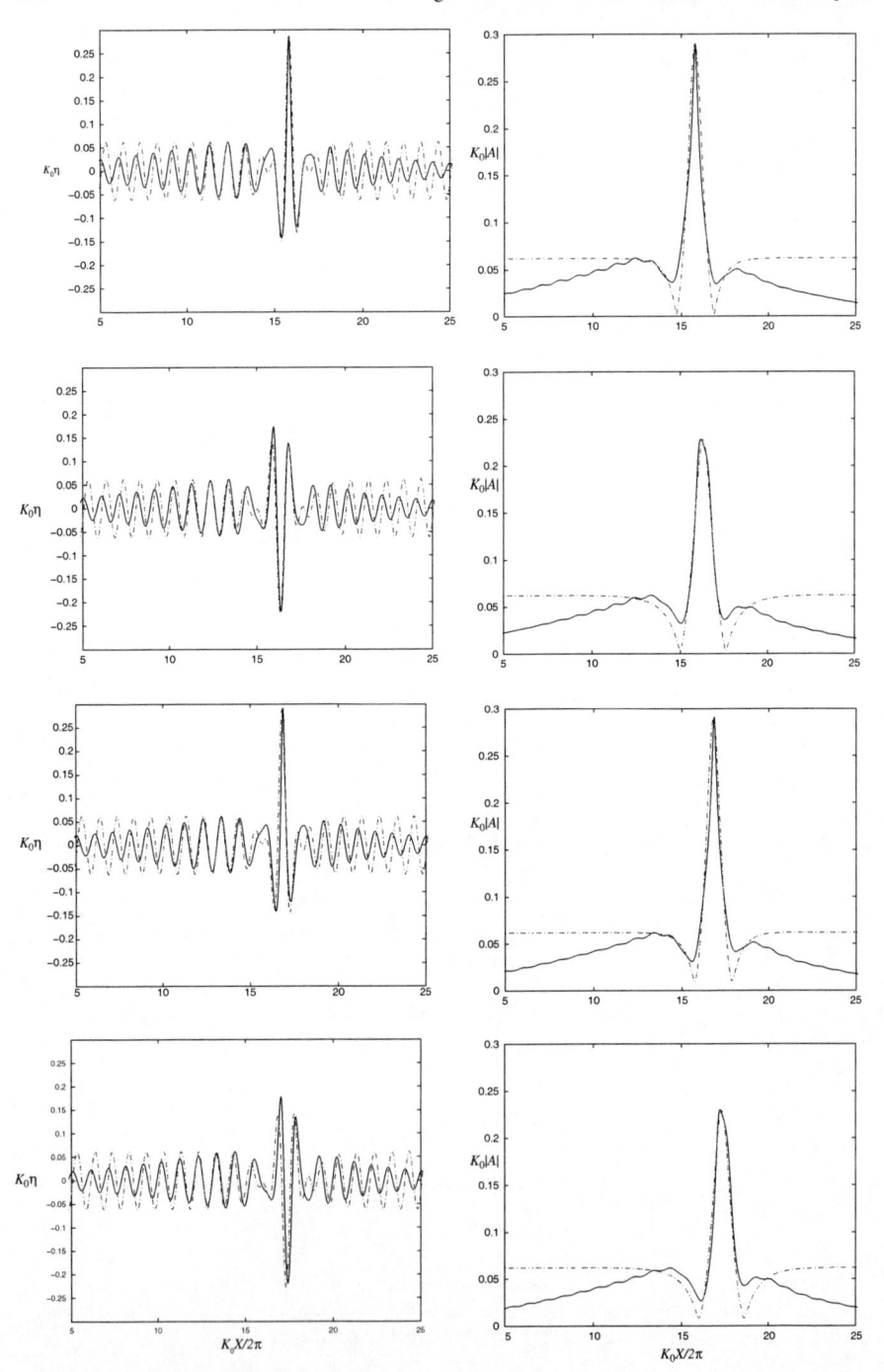

Fig. 4.19 Comparison of the fully nonlinear simulation (*solid*) and the fitted exact NLS solution (*dash*). Elevation (*panels on the left*) and envelope (*panels on the right*) of the surface elevation at $t/T_0 = 155, 156, 157, 158$ versus dimensionless coordinate. K_0 and T_0 denote the carrier wave number and period, respectivel. (see details in Clamond et al. 2006)

4.3 Rogue Wave Simulations within the Framework of the Fully Nonlinear Equations

Evolution and interaction of strongly nonlinear envelope solitary waves is considered in Zakharov et al. (2006b). The exact solutions of the NLS equation—namely, envelope solitons—have been used to initialize the computation. Weakly nonlinear wave packets behave similarly to the solutions of the NLS equation; they may propagate without deformation and preserve their identity rather well (Fig. 4.20a). Larger

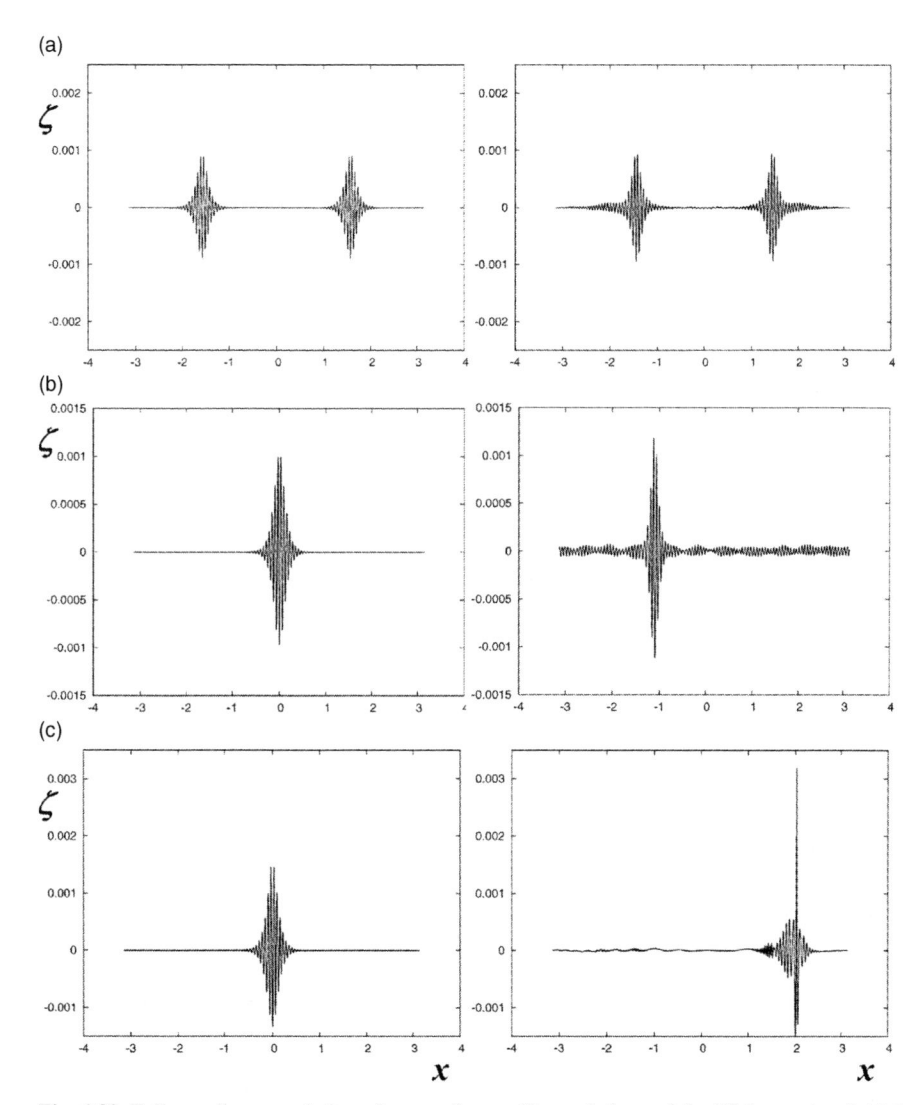

Fig. 4.20 Fully nonlinear evolution of an envelope soliton solutions of the NLS equation. Initial conditions are given in the *left panels*, the result is presented in the *right panels*. (**a**) Collision of two envelope solitons, each with steepness 0.085. (**b**) Evolution of an envelope soliton with steepness 0.1. (**c**) Evolution of an envelope soliton with initial steepness 0.14 (Zakharov et al. 2006b, reproduced with permission from Elsevier)

initial steepness results in modification of the profile of the envelope and radiation (Fig. 4.20b). The envelope approximation completely fails when the steepness is about 0.15 (Fig. 4.20c). The initial wave packet undergoes an additional compression, obviously related to strongly nonlinear effects, leading to the formation of a very high wave. This effect can be explained when the envelope solutions of a definite critical amplitude are unstable and can collapse.

4.4 Statistical Approach for Rogue Waves

It was shown by Caponi et al. (1982) and Yasuda and Mori (1997) that modulated water wave trains may evolve to chaotic states. This feature suggests the use of statistical and spectral descriptions. In looking at the sea surface, we are struck by both randomness and regularity of the wave field. Hence, the prediction of wave parameters of irregular waves may be achieved through a statistical approach.

For 1D propagation, Janssen (2003) studied the influence of the nonlinear four-wave interactions on the occurrence of large surface waves over deep water, using the Zakharov equations (Zakharov 1968, Krasitskii 1994) as a basis with both resonant and non-resonant interactions taken into account. The former interaction evolves on the characteristic time scale $(s^4\Omega)^{-1}$, whereas the latter has a much shorter characteristic time scale $(s^2\Omega)^{-1}$. At the same time, Dysthe et al. (2003) considered the stability of moderately narrow bell-shaped spectra by numerical simulation of the Dysthe equation. It was found that, regardless of the initial spectral bandwidth, the spectra evolve within the characteristic Benjamin-Feir time scale, $(s^2\Omega)^{-1}$, from a symmetric to an asymmetric shape, with a frequency downshifting of the peak. For 2D propagation, the computations of the latter authors confirm the $K^{-2/5}$ (or Ω^{-4}) power law of the spectrum in the inertial range. Using a truncated JONSWAP spectrum as initial conditions, and two kinds of angular distributions corresponding to short- and long-crested waves, respectively, Socquet-Juglard et al. (2005) found similar results and reported on the probability of the occurrence of rogue waves, too. For crest heights less than four times the standard deviation (very close to the significant wave height H_s), they showed that the distributions of surface elevation and crest height fit very well with the theoretical second-order distributions of Tayfun (1980). For larger waves (elevation higher than H_s), this is not always the case. For long-crested waves with a normalized spectral width $\Delta\Omega/\Omega$ less than the steepness s, an increase of the extreme wave events during a phase of spectral change is observed, whereas for short-crested waves, the spectral change does not seem to have much effect on the distribution of extreme wave events. To conclude this extreme wave analysis, Socquet-Juglard et al. (2005) found that the Tayfun distribution is a good approximation, even up to five standard deviations.

The key parameter controlling the importance of the nonlinear wave-wave interactions is the *Benjamin–Feir Index* (BFI) which is the ratio of the wave steepness to the spectral bandwidth. We define the BFI following Janssen (2003) as

$$I_{BF} = \sqrt{2}\frac{K\eta_{rms}}{\Delta\Omega/\Omega} = 2\sqrt{2}\frac{K\eta_{rms}}{\Delta K/K}, \quad (4.106)$$

4.4 Statistical Approach for Rogue Waves

where K and Ω are the mean wavenumber and frequency of the waves, and $\Delta\Omega/\Omega$ is the spectral bandwidth. One may find other possible definitions of the BFI at the end of this section (see also Olagnon and Magnusson 2004) for the collection of the BF indices applied for the wave record analysis. The dispersion in the deep-water case yields the relation $\Delta K/K = 2\Delta\Omega/\Omega$. The mean values and variances may be defined through spectral moments (see Sect. 2.2). The root mean-square surface displacement η_{rms} is related to the root mean-square amplitude A_{rms} via $A_{rms} = 2^{1/2}\eta_{rms}$. The wave amplitude is assumed to vary slowly compared with the carrier sinusoidal wave length. Therefore,

$$I_{BF} = \frac{KA_{rms}}{\Delta\Omega/\Omega} = 2\frac{KA_{rms}}{\Delta K/K}. \tag{4.107}$$

Following Alber (1978), the random wave field is stable when $I_{BF} < 1$ (here σ_r from Eq. (4.48) is equal to $\Delta\Omega$). In the opposite case, the BF instability is potentially possible if condition (4.39) is satisfied.

The BFI provides a convenient indicator for prediction of modulational instability. A number of recent research projects were aimed at establishing the relationship between the BFI and rogue wave-probability occurrence. Stochastic simulations of random wave fields and laboratory experiments have been performed, where the spectrum evolution and probability of extreme wave occurrence were compared against the values of the BFI (Onorato et al. 2001, 2004, 2005, 2006b; Janssen 2003; Dysthe et al. 2003; Socquet-Juglard et al. 2005). Under the assumptions of weakly non-Gaussian and narrow-band wave trains, Mori and Janssen (2006b) showed that the wave height and the maximum wave height probability distribution depend on the wave variance and kurtosis. The fourth-order statistical moment (kurtosis, κ) is a convenient parameter for measuring the non-Gaussianity of the wave field. For 1D propagation, it is found that the probability of occurrence of extreme wave events increases with kurtosis. The following support of this feature was derived by Mori and Janssen (2006b)

$$\kappa - 3 = \frac{\pi}{\sqrt{3}}I_{BF}^2 \tag{4.108}$$

(the Gaussian process corresponds to $\kappa = 3$). Hence, the kurtosis and the BFI are dependent parameters, and their growths lead to an increase of rogue wave occurrence. The relationship between the freak wave occurrence probability observed in numerical simulations and natural observations was discussed in Mori and Janssen (2006a).

A directional sea was considered in Onorato et al. (2002) within the framework of the extended Dysthe equation. Nonlinear interactions of codirectional waves lead both to an increase of the kurtosis and probability of occurrence of extreme waves, whereas for multidirectional waves the kurtosis is shown to oscillate around $\kappa \approx 3$, indicating that the probability density function for the wave amplitudes is approximately Gaussian.

Let us consider a slightly perturbed plane wave with amplitude A, mean wavenumber K, and long perturbation wavenumber defined by ΔK. According to the instability condition (4.39) for deep-water waves, the plane wave may be unstable if

$$2\sqrt{2}\frac{KA}{\Delta K/K} = \sqrt{2}\frac{KA}{\Delta\Omega/\Omega} > 1. \tag{4.109}$$

From a formal point of view, in terms of the BFI (4.107), this condition transforms into

$$\sqrt{2}I_{BF} > 1. \tag{4.110}$$

The wave modulations split the carrier into groups. The number of individual waves within such a group may be naturally defined as

$$n_x = \frac{K}{\Delta K}, \quad \text{and} \quad n_t = \frac{\Omega}{\Delta\Omega}, \tag{4.111}$$

where n_x and n_t are the numbers of individual waves observed in a snapshot and measured in one point time series, respectively. On deep water, they satisfy the condition $n_t = 2n_x$. Hence, definition (4.107) results in the dimensionless estimation

$$I_{BF} = \bar{s}n_t, \tag{4.112}$$

where $\bar{s}$ denotes the averaged steepness, $\bar{s} \equiv KA_{rms}$.

Taking into account the normalization (4.58), the "mass" integral (4.65) for one wave envelope may be written in the form

$$M \approx \pi 2\sqrt{2}KAn_x = \pi\sqrt{2}I_{BF}, \tag{4.113}$$

where the length of the envelope is estimated as $2\pi/\Delta K$. Hence, the soliton number (4.64) for a smooth pulse-like initial condition for the NLS equation is equal to

$$N_s = \left[\sqrt{2}I_{BF} + \frac{1}{2}\right]. \tag{4.114}$$

Relations (4.113) and (4.114) link the statistically defined BF index and the dynamical parameters of M and N_s. They are, roughly speaking, proportional to each other, when the 1D version of the NLS equation (4.33) is considered,

$$\frac{nonlinearity}{dispersion} \propto \frac{\Omega K^2 A^2}{2} \cdot \frac{8K^2}{\Omega(\Delta K)^2} = I_{BF}^2. \tag{4.115}$$

In this sense, the BFI is an analogue to the Ursell number, which is well known for shallow-water waves (see Chap. 5). Table 4.1 collects some important values of these parameters. It is seen that for different applications, the choice of different parameters may be convenient. When dealing with deterministic waves, it is more pertinent to use the quantity $\sqrt{2}I_{BF}$ (see Osborne et al. 2005, Slunyaev 2006).

In Onorato et al. (2001), Janssen (2003), Mori and Janssen (2006b), Gramstad and Trulsen (2007), and Tanaka (2007), it is shown through numerical experiments that the growth of the BFI index indeed qualitatively changes the statistical properties of the wave fields, but the change is not so abrupt and the threshold value of the index is not so obvious. The requirement of robust definition of this parameter

4.5 Laboratory Experiments of Dispersive Wave Trains with and without Wind

Table 4.1 Key values of parameters characterizing the nonlinear effects versus dispersion

Threshold	Parameter		
	BFI, I_{BF}	M/π	N_s
Rise of envelope solitons from a pulse-like packet	$\geq 1/\sqrt{8} \approx 0.35$	$\geq 1/2$	≥ 1
Onset of the plane wave modulational instability instability	$\geq 1/\sqrt{2} \approx 0.71$	≥ 1	≥ 1
One isolated envelope soliton	$1/\sqrt{2} \approx 0.71$	1	1
Cancellation of the BF instability instability due to randomness	< 1		

on the basis of real natural measurements, where noise perturbations always exist, opens a new problem (Olagnon and Magnusson 2004, 2005). See the discussion in Sect. 4.7.2.

4.5 Laboratory Experiments of Dispersive Wave Trains with and without Wind

Within the framework of dispersive focusing, Sect. 4.3.4 refers to experimental results conducted in the large wind-wave tank of IRPHE at Marseille, Luminy (see Fig. 4.21). The facility consists of a closed loop wind tunnel positioned above a water tank 40 m long, 1 m deep, and 2.6 m wide. The wind tunnel above the water flow is 40 m long, 3.2 m wide, and 1.6 m high. The blower can produce wind velocities up to 14 m/s, and a computer-controlled wavemaker submerged under the upstream beach can generate regular or random waves in a frequency range from 0.5 Hz to 2 Hz. Particular attention has been paid to simulating a pure logarithmic mean wind-velocity profile with constant shear layer over the water surface. A trolley installed in the test section allows probes to be located at different fetches all along the facility. The fetch is defined as the distance between the probes on the trolley and the end of the upstream beach where air flow meets the water surface. The water surface elevation is measured by using capacitive wave gauges: one is located at a fixed fetch 1 m from the upstream beach, and the others are installed on a trolley to measure the water surface elevation at different fetches from the upstream

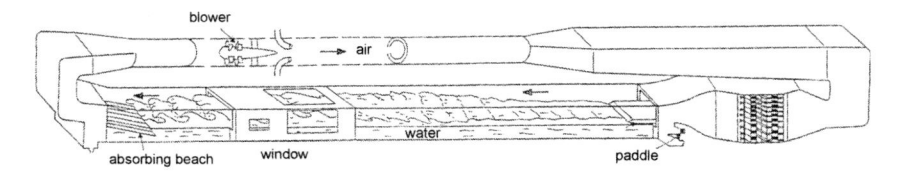

Fig. 4.21 A schematic description of the Large Air-Sea Interactions Facility, IRPHE

beach. The longitudinal and vertical air flow velocity fluctuations have been measured by means of an x-hot wire anemometer.

As in Sect. 4.3.4, extreme wave events are generated by means of a dispersive focusing mechanism with and without wind. The same initial wave train is generated and propagated without wind first, and under wind action for various values of the wind velocity afterwards. When the wind blows, the focusing wave train is generated once the wind waves have developed. For each value of the mean wind velocity U_w, the water surface elevation is measured at 1 m fetch and at different fetches between 3 m and 35 m. The wavemaker is driven by an analog electronic signal to produce this signal linearly varying with time from 1.3 Hz to 0.8 Hz in 10 s, with almost constant amplitude of the displacement. The wavemaker is totally submerged to avoid any perturbation of the air flow that could be induced by its displacement.

Figure 4.22 shows two time series of the probe located at 1 m fetch, recorded with no wind, and under a wind speed of $U_w = 6$ m/s. The probe record, corresponding to a wind velocity equal to 0 m/s, is artificially increased by 10 cm for more clarity of the figure. We see that the two signals are very similar. Some weak differences in amplitude are observed locally. Nevertheless, it is seen that no significant variations are observed, and the experiment is considered to be repeatable in the presence of wind.

More details on experiments conducted in the large wind-wave tank of IRPHE, can be found in Kharif et al. (2008). These results were anticipated in Sect. 3.3. Figure 3.11 presents the time series of the water surface elevation at different fetches for $U_w = 0$ m/s. For the sake of clarity, as it has been done for Fig. 4.22, the probe records given here are recursively increased by 10 cm. As predicted by the linear theory of free deep water waves (no wind), dispersion makes short waves propagate more slowly than long waves, and as a result, the waves focus at a given position in the wave tank leading to the occurrence of a large amplitude wave. Downstream from the point of focus, the amplitude of the group decreases rapidly (defocusing).

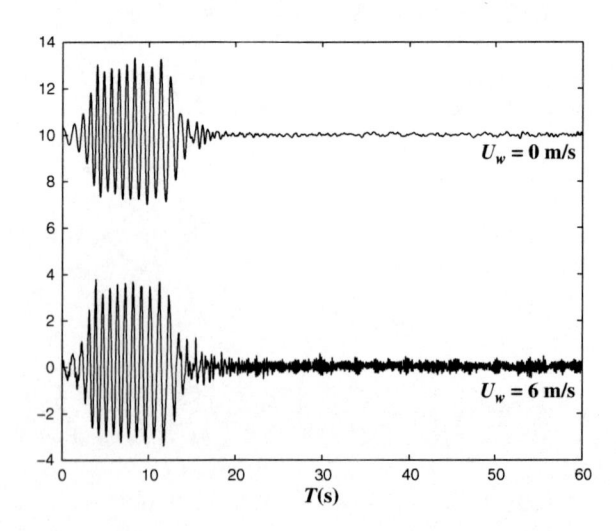

Fig. 4.22 Surface elevation (in cm) at fetch $X = 1$ m for wind speeds $U_w = 0$ and $U_w = 6$ m/s (note that for $U_w = 0$ m/s, the origin of the elevation corresponds to the value 10 cm)

4.5 Laboratory Experiments of Dispersive Wave Trains with and without Wind

Figure 3.12 shows the same time series of the water surface elevation, at several values of the fetch X, and for a wind speed $U_w = 6$ m/s. The wave groups, mechanically generated by the wavemaker, are identical to those used in the experiments without wind (see Fig. 4.22). Some differences appear in the time-space evolution of the focusing wave train. One can observe that the group of the extreme wave event is sustained longer.

Figure 4.23 gives the amplification factor as a function of the distance from the upstream beach for several values of the wind velocity, equal to 0 m/s, 4 m/s and 6 m/s. We can see that the effect of the wind is twofold: (i) it weakly increases the amplification factor; and (ii) it shifts the focus point downstream. Moreover, contrary to the case without wind, an asymmetry appears between focusing and defocusing stages. The slope of the curves corresponding to defocusing is modified. Note that before the focus point, the wind has no effect on the amplification factor. One can observe that the rogue wave criterion (I.1) is satisfied for a longer period of time. It is also interesting to emphasize that the rogue wave criterion is satisfied for a longer distance, while the wind velocity increases.

The numerical results obtained in Sect. 4.3.4 are confirmed by the experiments, at least qualitatively. A detailed physical analysis of wind-wave coupling over focusing groups may be found in Kharif et al. (2008).

Through experimentation, Baldock et al. (1996) investigated the spatio-temporal focusing of a large number of water waves at one point in space and time to produce a large transient wave group. The experiments were conducted in a 20 m long and 0.3 m wide wave flume. The facility has a maximum working depth of 0.7 m. The waves are generated by a flat bottom-hinged paddle located at one end of the wave flume. The period of the generated waves can vary from 0.4 s to 2.0 s. A total of six surface-piercing wave gauges were used to measure the surface elevation

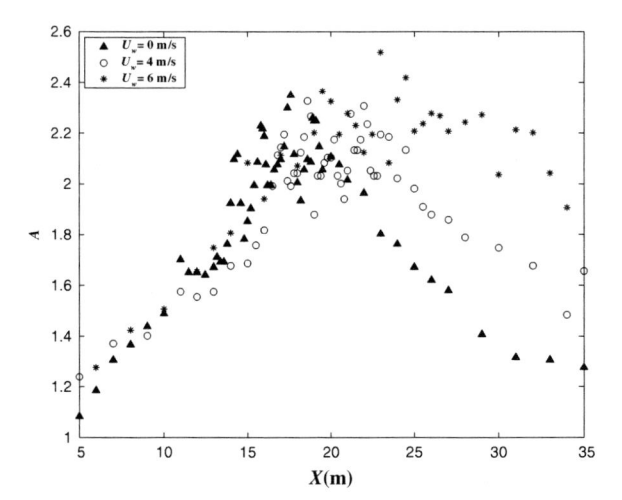

Fig. 4.23 Evolution of the amplification factor $A(X, U_w)$ as a function of the distance for several values of the wind speed

at fixed spatial locations. Baldock et al. (1996) adopted the approach developed by Rapp and Melville (1990) to create an extreme wave within the laboratory flume. They used a linear solution to determine the appropriate phasing of the various wave components. Owing to nonlinear wave-wave interactions present in the experiments, the theoretical and experimental focal points and focus times are different. To simplify the experimental procedure, they imposed that the focus point be located at a fixed distance down-stream of the paddle and the focusing time be set to zero. Measurements of the water surface elevation were compared with both linear wave theory and a second-order solution derived by Longuet-Higgins and Stewart (1960). The experimental results showed that the focusing wave mechanism produces the occurrence of an extreme wave event whose nonlinearity increases with the wave amplitude and reduces with increasing bandwidth. A comparison of the first- and second-order solutions shows that the wave-wave interactions generate a steeper envelope, in which the central wave crest is higher and narrower, whereas the adjacent wave troughs are broader and less deep. The authors suggested that the formation of a focused wave group involves a significant transfer of energy into both higher and lower harmonics.

Within the framework of 2D wave fields, Grue and Jensen (2006) reported velocity and acceleration fields in six very large wave events realized in a series of wave tank experiments. The wave slope is in the range 0.40–0.46 and exceeds the previously mentioned laboratory study of large waves (Baldock et al. 1996) by a factor of about 50%. Focusing water waves were produced in a 24.6 m long wave tank in the Hydrodynamic Laboratory at the University of Oslo. The tank width is 0.5 m and the water depth 0.72 m. The velocities and the material acceleration fields of the waves are obtained by employing an extended Particle Image Velocity (PIV) system (see Jensen et al. 2001). The velocity vector has a magnitude comparable to the wave speed in the strongest case, and is manifested in the jet that develops at the front face of the breaking waves. The nonbreaking waves present a maximal horizontal acceleration up to about $0.70\,g$ in the front face of the wave at vertical level about halfway to the crest. The overturning events present horizontal accelerations up to $1.1\,g$ and vertical accelerations up to $1.5\,g$ in the front face of the wave, at the base below the overturning jet.

Onorato et al. (2006b) conducted a series of experiments in a long water-wave flume at Marintek in Trondheim (Norway). The length of the tank is 270 m, its width is 10.5 m, and its depth is 10 m for the first 85 m, then 5 m for the rest of the flume. A horizontally double-hinged flap type wavemaker located at one end of the tank was used to generate the long-crested waves, whereas an absorbing beach is located at the end opposite from the wavemaker. Several probes were used along the tank to measure the wave surface elevation. Three experiments corresponding to three different JONSWAP spectra with different values of the Phillips parameter α and the peakedness γ (see (2.116)) were conducted. The main goal of these experiments was to give experimental support to the results of theoretical and numerical studies developed previously. According to these studies, it was suggested that the modulational instability was responsible for the occurrence of extreme wave events. The modulational instability or the Benjamin-Feir instability that was obtained for uniform wave trains within the framework of deterministic approaches is assumed to work in random wave fields, too. Onorato et al. (2006b) showed that for long-crested water waves and large values of the Benjamin-Feir index, the second-order theory is not relevant to describe the tails of the probability density function of wave crests and wave heights. They showed that the probability of finding an extreme wave was underestimated by more than one order of magnitude if second order theory is considered, and found that the deviation was due to the modulational instability mechanism occurring for large BFI.

4.6 Three-Dimensional Rogue Waves

Until now, we have mainly paid attention to 2D aspect of the rogue wave formation. In this section, 3D aspects are discussed.

Rogue waves in the form of "walls of water" (see Fig. 1.2b) may potentially be described within the framework of 2D models (i.e., unidirectional wave propagation). At the same time, transversal effects are known to be important—for example, the NLS envelope soliton is transversally unstable. The localized "pyramidal" waves, like those in Fig. 1.2a,c, undoubtedly require consideration of the transverse wave direction. The geometrical focusing phenomenon may result from spreading waves. It is a linear mechanism of wave-energy focusing that was considered in Sect. 3.1. This primitive mechanism may be quite important in the real ocean, since papers report about higher probability of rogue wave occurrence in mixed seas.

Dispersive focusing is still efficient in 3D situations; and this kind of wave compression may be further enhanced by geometrical focusing (see Slunyaev et al. 2002). This results in more rapid and significant wave growth compared to the 2D case. If the dispersive wave train is far from the modulational instability threshold, the dispersive focusing prevails similarly to the linear case. The presence of random wave components may hide the deterministic process of rogue wave generation, but does not prevent the quasi-linear wave focusing as shown in Fig. 4.24. The rogue wave appears "from nowhere" and disappears at once.

Realistic fully nonlinear 3D simulations of directional wave focusing were conducted by Fochesato et al. (2007) (see Fig. 4.25). They found that the vertical 2D longitudinal cross section through an extreme wave crest looks quite similar to the characteristic shape frequently observed for rogue waves in the ocean: a tall and steep doubly asymmetric wave crest occurs in between two shallower troughs. The 3D wave generation yields a curved wave front before focusing occurs. A shallow circular trough forms in front of the focused wave ("hole in the sea"), followed by a deeper trough with a crescent shape. For a small time prior to breaking, the 3D shape of the focused wave appears to be almost pyramidal. By contrast, during the focusing phase, as well as the development of overturning, the transverse shape of the wave through the crest tends to have a more rounded shape. The problem of reproducing the desired 3D wave shapes in tanks was investigated by Bonnefoy et al. (2005).

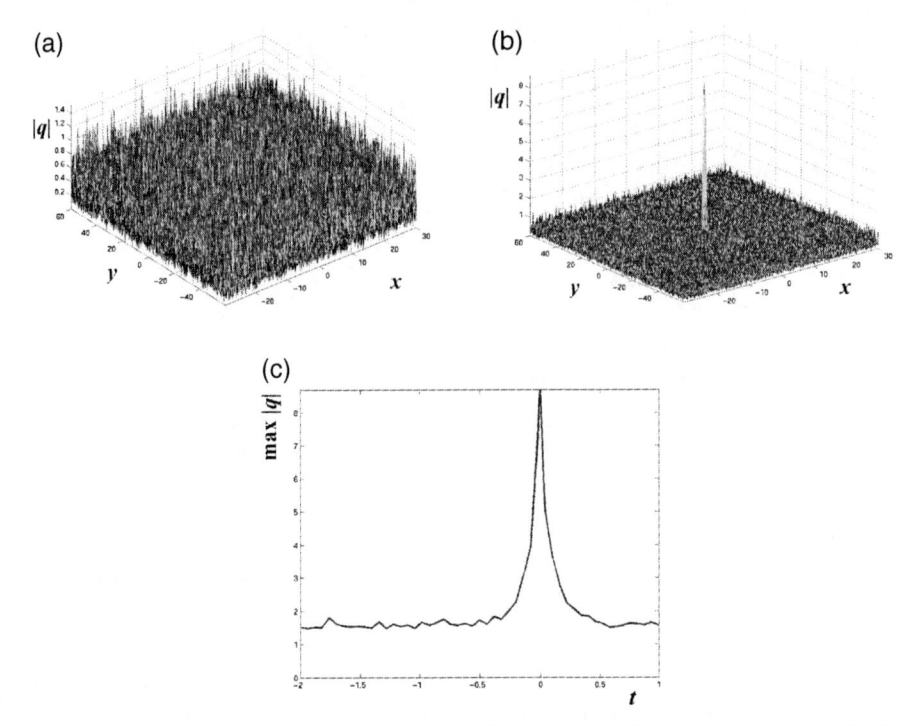

Fig. 4.24 Three-dimensional dispersive focusing of a wave train with modulated wavenumber in the presence of strong random wave components: (**a**) initial wave envelope, (**b**) moment of focusing, (**c**) record of maximum envelope amplitude versus time. Simulation within the framework of the NLS equation (see details in Pelinovsky et al. 2003)

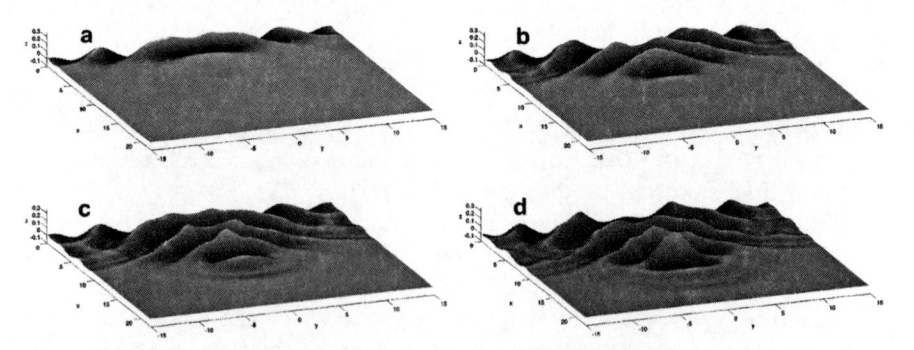

Fig. 4.25 Snapshots of 3D free surface evolution computed by Fochesato et al. (2007). The focused wave is starting to overturn in panel (**d**) (Reproduced with permission from Elsevier)

Johannessen and Swan (2001) extended the experimental investigation of Baldock et al. (1996) in 3D wave fields. They considered a laboratory study in which a large number of water waves of varying frequency and propagating in different directions, were focused at one point in space and time to generate a large wave event. Experiments were conducted in a basin located at Edinburg University. This facility has

4.6 Three-Dimensional Rogue Waves

a plan area of 24 m × 11 m and supports a constant working depth of 1.2 m. The water waves are generated by 75 numerically controlled wave paddles located along one of the longer sides of the wave tank. At the opposite end of the basin, a set of passive absorbers dissipates the incident wave energy. Different directional distributions are applied to the frequency spectra of the surface elevation. Johannessen and Swan (2001) showed that the directionality may have a profound impact effect upon the nonlinearity of a large wave event. When the sum of the wave amplitudes generated at the wave paddles is kept constant, an increase in the directional spread of the wave field results in lower maximum crest elevations. Conversely, when the generated wave amplitudes are increased until the onset of wave breaking, an increase in the directional spread allows larger extreme waves to evolve. The authors suggested that these results are due to the redistribution of the wave energy within the frequency domain. They emphasized the rapid widening of the free-wave regime in the vicinity of an extreme wave event, too.

In 2D (*XZ*) geometries, the modulational instability is strongly associated with the solitary solutions of the NLS equation (breathers or homoclinic orbits). These objects are conserved during the evolution due to the integrability property of the NLS equation. The 3D version of the NLS equation, as well as the Davey-Stewartson system[5] are nonintegrable. Therefore, the wave dynamics are more complicated for comprehension. For instance, the wave field, growing due to geometric or dispersive grouping but initially stable with respect to modulational instability, may then pass the threshold of nonlinear self-focusing and continue further enhancing due to nonlinearity. The Benjamin-Feir instability diagram (Fig. 4.1) provides a rich variety of unstable growing wave packets. Some shapes of 3D rogue waves spawned by modulational instability have been presented in the papers by Osborne et al. (2000) and Slunyaev et al. (2002), respectively. As an example, the 3D rogue wave given in Fig. 4.26d is more than seven times amplified with respect to the initial weakly modulated waves. The quasi 2D modulational instability (Fig. 4.26a,b) is followed by the strictly 3D modulational dynamics (Fig. 4.26c,d), which results in the formation of a huge wave isolated in both longitudinal and transversal directions. It is readily seen from Fig. 4.26e that the 3D rogue wave growth ($t \approx 4.1$) is more sudden and significant than the 2D dynamics ($t \approx 3$).

In water of infinite depth, it is well known that the 2D modulational instability is dominant for small to moderate initial steepness and evolves into a recurrence phenomenon (the Fermi-Pasta-Ulam recurrence) for small initial wave steepness (see Sect. 4.1.1). Another kind of disturbance suffered from 3D instabilities (see McLean 1982a,b) exists and becomes dominant for larger values of the steepness. This instability may lead to the formation of horseshoe patterns evolving into 3D spilling breakers. These three-dimensional patterns take the form of crescent-shaped perturbations riding on the basic waves. Three-dimensional horseshoe patterns were observed in experiments of Su et al. (1982) and Su (1982), Melville (1982), Kusuba and Mitsuyasu (1986, in presence of wind), and others.

---

[5] The DS system becomes integrable only in the shallow-water limit. In this case it does not show modulational instability.

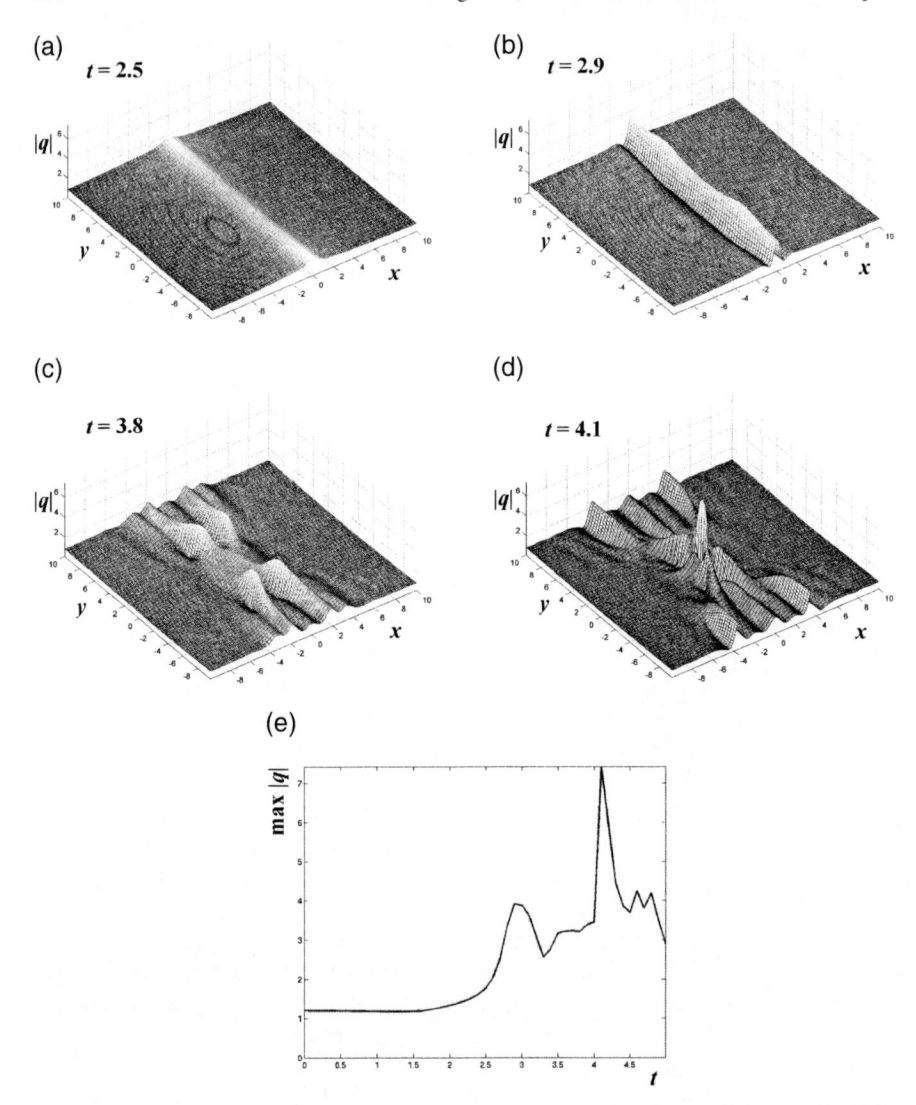

Fig. 4.26 Evolution of a weakly amplitude modulated plane wave within the framework of the NLS equation (see details in Slunyaev et al. 2002). Four snapshots of the envelope evolution (**a–d**) and the record of the maximum envelope amplitude versus time (**e**)

Generally, the two kinds of instability, namely the modulational instability and the crescent patterns that belong to class I and class II, respectively, coexist in the wave field. Depending on parameters such as the wave steepness of the initial Stokes wave and water depth, one can expect a competition to occur between the two classes of instability. Figure 4.27 illustrates the critical steepness together with the distinction between class I and class II dominances at the same depths. In the finite depth case, class II dominates in a large range of steepness and recurrence is

4.6 Three-Dimensional Rogue Waves

Fig. 4.27 Threshold steepnesses between class I and class II predominances and between class II recurrence and breaking: (**a**) the infinite depth case, (**b**) the finite depth case $K_0 D = 1$

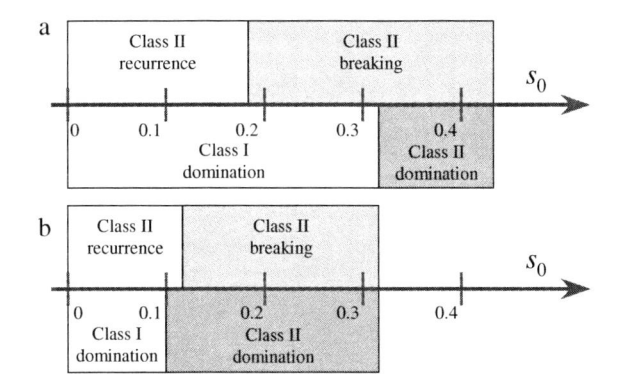

possible within this range. Note that for shallow water cases and relatively moderate steepness, instability of a plane Stokes wave is dominated by class II (Francius and Kharif 2006).

Numerical simulations by Fructus et al. (2005) and Kristiansen et al. (2005), taking into account both class I and class II instabilities, showed that for moderately steep waves ($s > 0.12$), their nonlinear coupling (involving the fundamental of the Stokes wave) results in breaking of the wave when in the initial condition only the modulational instability was considered. Furthermore, the breaking can occur for $s = 0.10$ when the initial unstable perturbation corresponds to the phase-locked crescent-shaped patterns. At the maximum amplitude of this instability, the modulational instability is excited followed by the breaking of the wave. For steeper waves, the strength of class II instability alone is sufficient to trigger the breaking of the wave. The nonlinear dynamics of the most unstable class II perturbation leads to breaking when $s > 0.17$ (see Fig. 4.27a).

Annenkov and Badulin (2001) selected the specific component peculiar to five-wave interactions in the frequency spectrum of the 20 min New Year Wave record. This component corresponds to class II instabilities phase-locked to the dominant component of the spectrum. In order to have a better understanding of the role of this kind of resonance in the formation of rogue waves, the authors performed numerical simulations of the Zakharov equation, which takes into account the modulational (which is a four-wave interaction) and five-wave interactions. Annenkov and Badulin (2001) showed that the cooperative effects of these interactions might be responsible for the occurrence of rogue waves and emphasize the role of oblique waves in this process.

Ruban (2007) investigated a weakly 3D evolution of modulationaly unstable wave patterns by means of fully nonlinear simulation and observed "zigzag patterns" with extreme waves in their turns formed during instability development. Recurrent dynamics of 3D wave patterns over deep and finite depth were simulated in recent papers (Kristiansen et al. 2005, Fructus et al. 2005) and are shown in Figs. 4.28 and 4.29, respectively. In the real sea, the hydrodynamic instability appears at the center of the crescent patterns when the wave steepness is above a threshold value.

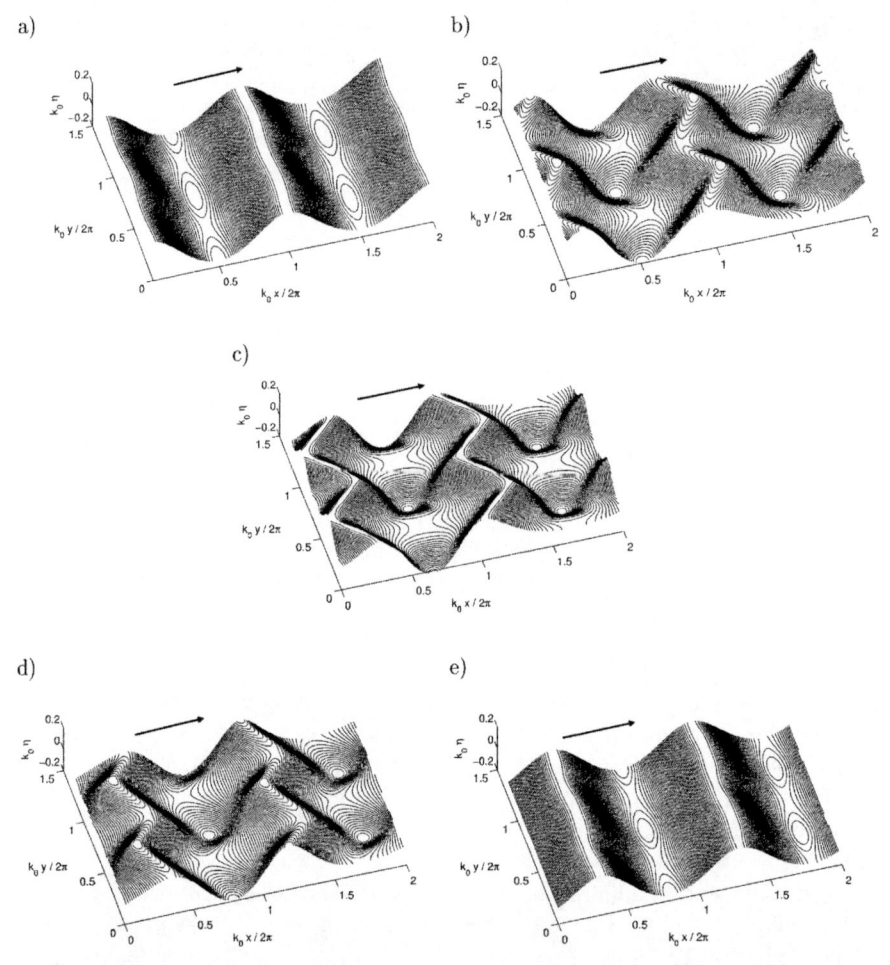

Fig. 4.28 Temporal evolution of the surface elevation during a recurrence cycle, fully nonlinear simulation of Kristiansen et al. (2005). The infinite depth case (Reproduced with permission from American Institute of Physics)

Recent findings of Gramstad and Trulsen (2007) by means of numerical simulation of the extended Dysthe equation show a conspicuous qualitative difference between the extreme wave dynamics in long- and short-crested seas. The paper reports about weak deviation of extreme waves from the Gaussian statistics when short crest lengths are concerned. On the other hand, the long crest wave statistics of freak waves is strongly non-Gaussian, and the Benjamin-Feir instability seems responsible for rogue wave formation. These results qualitatively agree with the predictions of Onorato et al. (2002, with extended Dysthe equations), Shukla et al. (2006, with coupled NLS equations) and Gibson et al. (2007) but appear to be conflicting with the studies of unstable crested waves by Onorato et al. (2006a, with coupled NLS

4.6 Three-Dimensional Rogue Waves

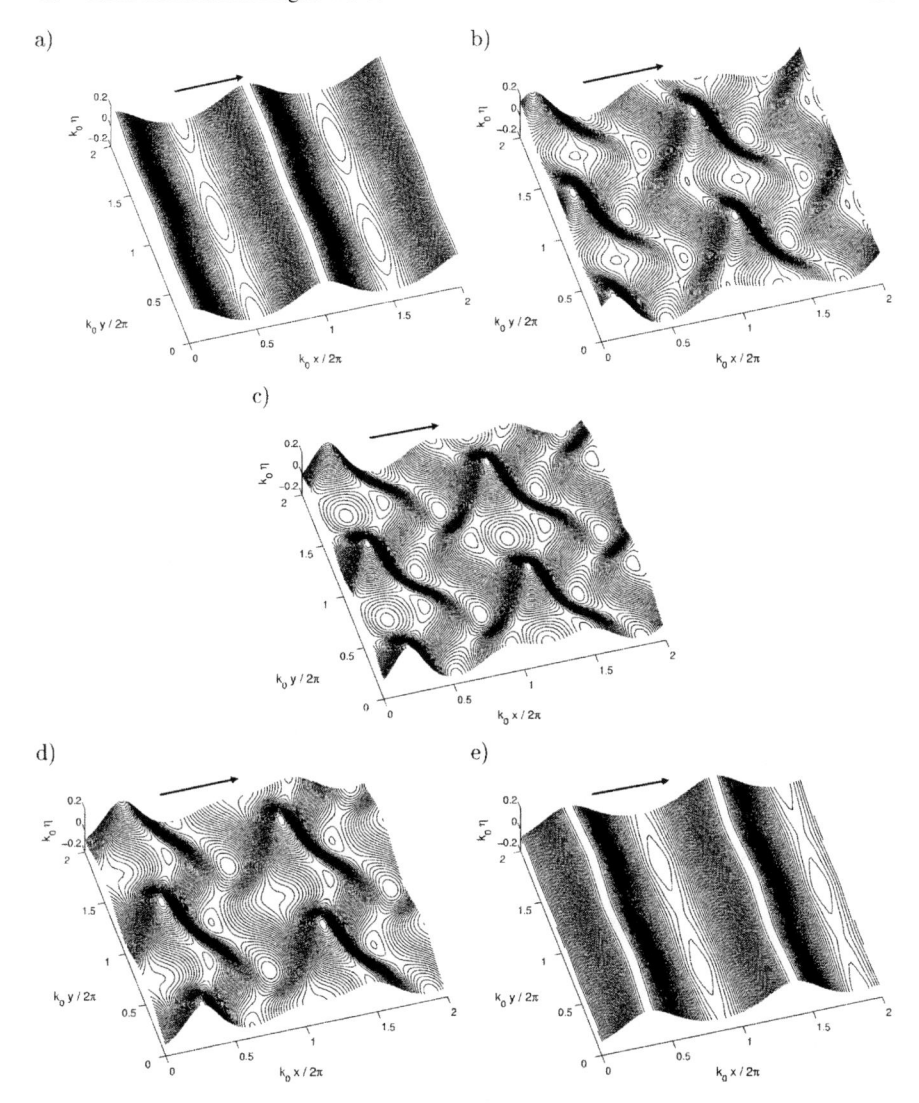

Fig. 4.29 Temporal evolution of the surface elevation during a recurrence cycle, fully nonlinear simulation of Kristiansen et al. (2005). The finite depth case $K_0 D = 1$

equations), fully nonlinear simulations of Ducrozet et al. (2007), and some natural observations (Pinho et al. 2004, Scott et al. 2005).

Although the real ocean is not homogeneous nor stationary, it was suggested by Haver (2005) and Gibson et al. (2007) that in seas of short-crested waves, some long-crested sub areas may exist in principle, which provides conditions for the high probability of the rogue wave occurrence.

Using the data collected from 1995 to 1999 by Lloyd's Marine Information Service, Toffoli et al. (2004) showed that a large percentage of ship accidents due to bad

Fig. 4.30 Pyramidal waves observed by Kimmoun et al. (1999) in a laboratory tank

weather conditions occurred in crossing sea states. Crossing sea sates are characterized by two dominant spectral peaks, and may be due to the interaction between a swell and a wind-wave sea coming from a different direction. This feature was also observed in the New Year Wave record. Onorato et al. (2006a) considered a weakly nonlinear model that describes the interaction of two-wave systems in deep water with two different directions of propagation. Under the assumption of narrow-band wave fields, they derived two coupled NLS equations from the Zakharov equation. As a main result, they showed that given a single unstable plane wave, the introduction of a second plane wave traveling in a different direction can increase the instability growth rates and enlarge the instability region. From their simple model, they suggested that the modulational instability could explain the formation of rogue wave events in crossing sea states. For more details concerning the stability of short-crested gravity waves due to the nonlinear interaction between two plane waves propagating in two different directions, see the papers by Ioualalen and Kharif (1994) and Badulin et al. (1995). These numerical and theoretical investigations on short-crested waves were followed by an experimental study conducted by Kimmoun et al. (1999) who observed pyramidal waves (see Fig. 4.30).

4.7 In Situ Rogue Waves

A great deal of theoretical investigations aimed at solving the rogue-wave phenomenon has been undertaken. Although some of the suggested physical mechanisms explain the occurrence of rogue waves rather well, the natural mechanisms that spawn rogue waves observed in the real ocean still need investigation. Instrumental measurements are the best source of getting information about real sea wave dynamics. The state-of-art in instrumental registrations is discussed in Sect. 1.2. Many registrations correspond to the case of deep or moderately deep water, but still are made in very different conditions throughout long periods of time. This makes direct statistical analysis quite hard or impossible. Some recent results of these statistical studies are given in Sect. 4.7.2. Section 4.7.1 is devoted to the analysis of the instrumental records themselves, trying to find out most possible information from "traces" of rogue waves.

4.7.1 Nonlinear Analysis of Measured Rogue Wave Time Series

4.7.1.1 Local Parameters

Local wave parameters may be used to reveal peculiar properties of measured rogue waves within the field of usual oceanic waves. To do this, shorter overlapping time intervals are extracted from the record. This procedure is known as Gabor or "windowing" transform. To reduce possible spurious effects due to the discontinuity of the time series at the boundaries, the Hanning data mask may be applied (Massel 1996). Examples of some local parameter estimations are given in Figs. 4.31 and 4.32 for two time series measured at the North Alwyn platform in the North Sea (see details in Slunyaev et al. 2005, Slunyaev 2006). The platform conditions correspond to sufficiently deep water ($KD > 3.6$), therefore we will restrict ourselves to the infinite depth approximation.

The mean frequency Ω is obtained as the spectral moment (2.109), Ω_p. The unfavorable result of considering a shorter time series gives a worse accuracy in statistical estimations and in particular the spectrum and all spectral parameters. The carrier frequency curves are given in Figs. 4.31 and 4.32 on panels A and B (the solid white line on the background of the Fourier time-frequency spectra) for two different durations of the sampling window T_{win}. One can observe some variations during the 20-min record, which become more evident if expressed in terms of group velocity C_{gr} (see Figs. 4.31D and 4.32D). The group velocity is obtained through the linear dispersion relation, since the measurement is available in only one point. The deviation of the group velocities observed in these cases is about 50%; it leads to the energy exchange between the individual waves. This may provide the wave growth or decrease and represents the simplest case of dispersive focusing. For a simple analysis of this process, the kinematic theory (3.13) may be used accompanied by the energy balance equation (3.5) (see Chap. 3):

$$\frac{\partial C_{gr}}{\partial T} + C_{gr}\frac{\partial C_{gr}}{\partial X} = 0, \qquad \frac{\partial \eta^2}{\partial T} + \frac{\partial}{\partial X}\left(C_{gr}\eta^2\right) = 0 \qquad (4.116)$$

where $\eta(X,T)$ is the surface elevation.

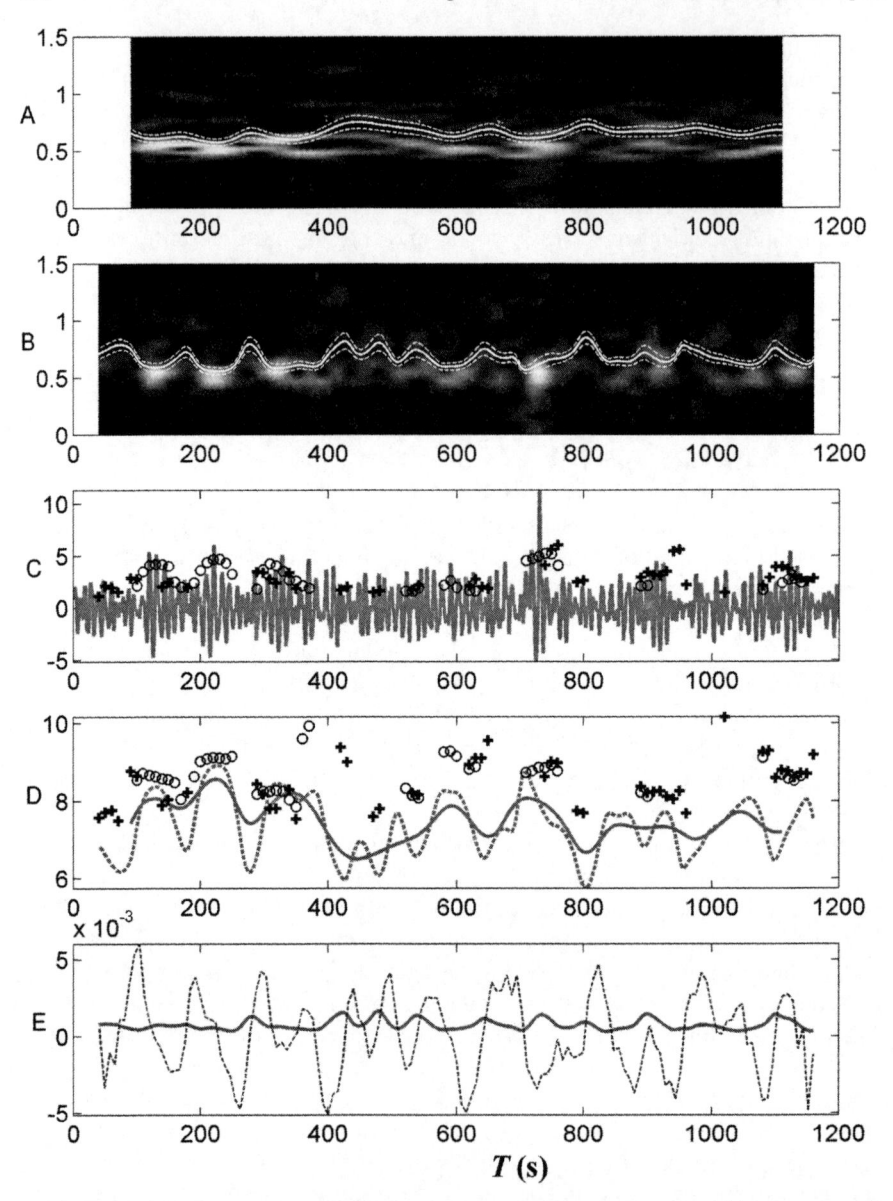

Fig. 4.31 Wave record made at the North Alwyn platform on November 18th, 1997 at 01:10. (**A**) Time-Frequency Fourier spectrum built for the sampling window of 117 s duration (about 10 wave periods); solid line shows the local mean frequency Ω, dashed lines bound the domain of Benjamin-Feir instability $\Omega \pm \Delta\Omega_{BF}$. **B**) The same as on panel A, but for the sampling window of 36 s duration (3 wave periods). **C**) Measured time series of the surface displacement (in meters). Symbols denote the determined amplitudes of solitary waves with permanent normalizing (*circles*) and flexible normalizing (*crosses*). **D**) Local group velocities (in m/s) defined for the sampling window of 117 s (*solid line*) and 36 s (*dashes*). Symbols denote the determined velocities of solitary waves: permanent normalizing (*circles*) and flexible normalizing (*crosses*). **E**) Growth rates σ_{BFmax} (*solid*) and σ_{dis} (*dashed*) (in s^{-1}) defined with the sampling window of 36 s

4.7 In Situ Rogue Waves 157

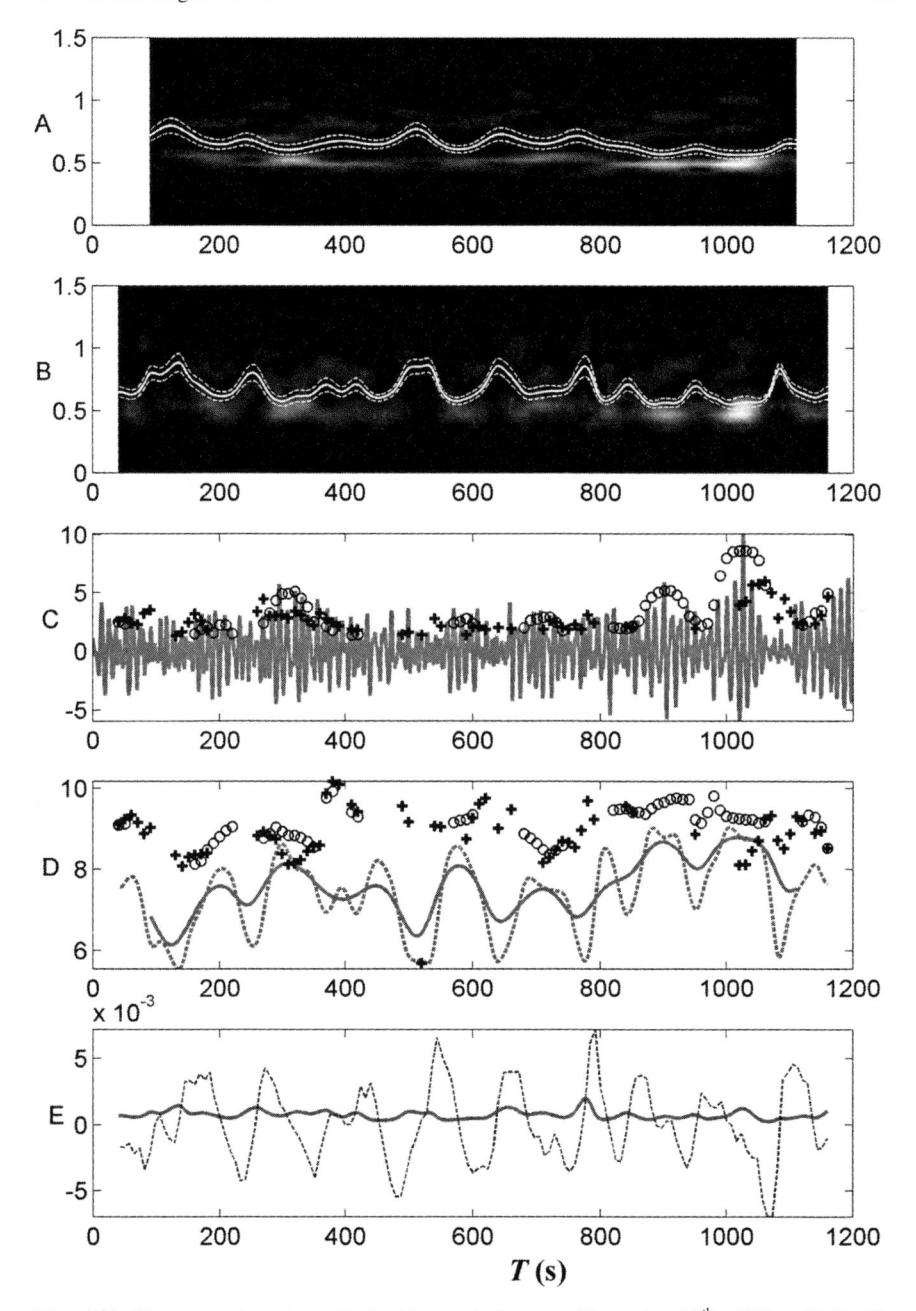

Fig. 4.32 Wave record made at North Alwyn platform on November 19th 1997 at 20:11. The legend is same as in Fig. 4.31

Then the total derivative of the energy quantity is given by:

$$\frac{d\eta^2}{dT} = 2\sigma_{dis}\eta^2, \quad \text{where} \quad \sigma_{dis} = \frac{1}{2C_{gr}}\frac{\partial C_{gr}}{\partial T}. \tag{4.117}$$

Parameter σ_{dis} expresses the exponential growth rate due to dispersive wave convergence. Waves grow when $\sigma_{dis} > 0$ and decay for $\sigma_{dis} < 0$. The dispersive growth rates computed for the time series are given in Figs. 4.31 and 4.32 (panel E, dashed lines).

4.7.1.2 Application of the IST Approach

The simple nonlinear theory is based on the NLS equation (4.35) under the extra assumption of unidirectional wave propagation. The spatial version of the dimensionless NLS equation has the form

$$iq_x + q_{tt} + 2q|q|^2 = 0, \tag{4.118}$$

where

$$t = \Omega_0 T - 2K_0 X, \ x = K_0 X, \ q = \frac{1}{\sqrt{2}} K_0 A^*, \tag{4.119}$$

and $A \equiv \eta_{01}$ is the complex envelope amplitude (see Eqs. (4.11), (4.13)). The spectral areas that are unstable with respect to long perturbations of the uniform Stokes waves may be estimated as the domain $(\Omega - \Delta\Omega_{BF}, \Omega + \Delta\Omega_{BF})$, where $\Delta\Omega_{BF}$ is defined with the help of the instability criterion (4.39) and deep-water dispersion relation as

$$2\frac{\Delta\Omega_{BF}}{\Omega} = \frac{\Delta K_{BF}}{K} < 2\sqrt{2}K\eta. \tag{4.120}$$

The unstable frequency domain $\Omega_0 \pm \Delta\Omega_{BF}$ is bounded by the dashed lines in Figs. 4.31 and 4.32 (panels A, B). The initial stage of the modulational growth is described by the exponential law with a maximum growth rate given by formula (4.41), which is, in the deep-water case,

$$\sigma_{BF\,\text{max}} = \frac{1}{2}\Omega K^2 \eta^2. \tag{4.121}$$

The two growth rates σ_{dis} and σ_{BFmax} (see Figs. 4.31 and 4.32, panels E) are used for rough estimates of the time scales of dispersive and nonlinear wave focusing effects. It is seen from the figures that dispersion typically works faster, while estimated modulational growth should take more than 500 s.

The nonlinearity of individual waves is characterized by the steepness, although the strength of self-focusing is characterized by another nonlinear parameter, which is the soliton number or the BFI (see Sects. 4.2 and 4.4). The "dynamical" definition of BFI (4.112) includes the number of individual waves observed in the wave group n_t. The number of waves within a packet is actually a convenient dimensionless parameter, and is often used for estimations.

4.7 In Situ Rogue Waves

Since the modulational instability occurrence is related to the homoclinic orbits, "unstable modes," or envelope solitons (see Sect. 4.2), a more accurate way to estimate the features of the modulational instability may be suggested by employing the concept of the envelope soliton. The envelope soliton of the NLS equation may be considered as the first approximation for oceanic solitary wave groups. Results reported in Sect. 4.3.5 concerning the steep NLS soliton-like envelopes, justify the adequacy of the quasi-soliton concept even in strongly nonlinear cases. For the equation in the form (4.118), the envelope soliton solution (4.60) is rewritten as

$$q_{es}(x,t) = A_{es} \frac{\exp\left(\frac{i(t-t_0)}{2V_{es}} - ix\left(\frac{1}{4V_{es}^2} - A_{es}^2\right) + i\theta_0\right)}{\cosh\frac{A_{es}}{V_{es}}(x - x_0 - V_{es}(t-t_0))}, \quad (4.122)$$

where the parameters t_0 and θ_0 are explicitly introduced, which are the time shift at position $x = 0$, and the initial phase. In Eq. (4.122), A_{es} and V_{es} are a dimensionless amplitude and velocity of the envelope soliton, respectively. The physical parameters, the amplitude of the wave packet A_{wp}, and its velocity V_{wp} are expressed as

$$A_{wp} = \frac{\sqrt{2}}{K_0} A_{es}, \quad V_{wp} = \frac{\Omega_0}{K_0(2 + V_{es}^{-1})}. \quad (4.123)$$

The applicability of the NLS theory (spectral narrowness) requires the quantity $|V_{es}^{-1}|$ being small.

When envelope solitons interact with other waves, the dynamics of the wave field may become complex. The possibility of detecting hidden solitons in time series may provide an effective tool in understanding and predicting nonlinear wave dynamics. This can be done with the help of the Inverse Scattering Technique (see Sect. 4.2.1). The spectrum of the scattering problem is time independent, and its discrete part corresponds to envelope solitons. Let us consider the scattering problem (4.59) for the infinite line; then the soliton parameters are simply related to the spectrum as follows

$$A_{es} = 2\,\mathrm{Re}\lambda \quad \text{and} \quad V_{es}^{-1} = 4\,\mathrm{Im}\lambda \quad (4.124)$$

instead of (4.67). The complete solution of the inverse scattering problem for function $q(x = 0, t)$—i.e. determination of t_0 and θ_0—requires knowledge of the eigenmodes. The parameter t_0, which defines the position of the envelope soliton in the time series, may be well localized if short overlapping extracts from the time series are considered (employing the windowing transform). Thus, the direct scattering problem is solved in a sliding sampling window of length t_{win} that identifies the position t_0 of solitons. If wave groups of large amplitude are of interest, the window t_{win} is bounded owing to the conservation of the mass parameter M_{es} (4.69) for the envelope solitons (i.e., steep solitons are narrow).

It is necessary to define the carrier wave frequency when considering the NLS equation (4.118). Panels A in Figs. 4.31 and 4.32 show its variation. Therefore, to follow the variation of the frequency, a short window should be used. On the other hand, it is more difficult to obtain a reliable estimate of this value within a short window, preserving only few wave periods. The number of envelope solitons is governed by the mass parameter, as Eq. (4.64). In the case of the spatial version of the NLS equation (4.118), the following estimate may be done

$$N_s \propto K_0 \Omega_0 \propto \Omega_0^3, \qquad (4.125)$$

where the deep-water dispersion relation is used. Therefore, accurate determination of the carrier wave frequency may be crucial for this method.

The soliton amplitudes that have been obtained with the help of this approach are plotted as circles and crosses on panel C of Figs. 4.31 and 4.32. The corresponding soliton velocities are given on panel D. The mean frequency is defined via two methods. First, it is defined as the spectral moment Ω_p (2.109) of the whole 20-min record ("permanent normalization"), and second, as the spectral moment of each short extract ("flexible normalization"). These cases correspond to circles and crosses in the figures. It is evidently seen that sometimes the results are rather different. After having a look at the curves of the group velocity (panel D), it becomes clear that a soliton vanishes if the mean group velocity increases. The effect of non-uniformity on modulated wave packets was considered by Duin (1999) with the same qualitative conclusion: the BF instability is depressed when the local group velocity increases and is intensified when C_{gr} becomes smaller.

Only the first (steeper) solitons defined in extracts are shown in the figures. Other solitons are usually much smaller and assumed not to be very trustworthy. Although the found solitons can often be seen by eye, they interact nonlinearly with other waves, and in other conditions may be hidden by the surrounding waves.

The idea to seek solitons in a time series was, evidently, first realized by Osborne and Petti (1994) for the shallow-water case, when the waves were described within the framework of the Korteweg-de Vries equation. Recently, a similar technique has been used for the study of freak waves over deep water within the NLS approach (Osborne et al. 2005, Islas and Schober 2005, Schober and Calini 2008). In contrast to the previous description, they suggest the use of periodic domains and the determination of the eigenmodes (full reconstruction of unstable modes). This makes the approach more difficult when employing the theta-functions, whereas applying the infinite line scattering problem formulation admits the description of wave groups with the help of breathing solutions considered in Sect. 4.2.3.

To estimate the contribution of the solitary part in the observed freak waves, let us assume a rogue wave is the result of the interaction of an envelope soliton with a plane wave. Then the "solitary part" is defined as A_{wp}/A_{fr}, where A_{wp} is the detected amplitude of the soliton (4.123), and A_{fr} is the Hilbert envelope amplitude including the freak wave obtained directly from the time series. The contribution of the background waves is estimated as $H_s/(2A_{fr})$. According to the analysis provided in Sect. 4.2.3, these contributions linearly supplement each other as Eq. (4.75). They are represented by the solid and hatched areas in Fig. 4.33, respectively. Eleven analyzed rogue waves measured at oil platforms in the North Sea are used in the figure. The rest (the empty areas) estimates the effects that are not taken into account. It may be noticed that the first two contributions (the solitary part and the significant background) may often completely explain the registered wave amplitude; this obvi-

4.7 In Situ Rogue Waves

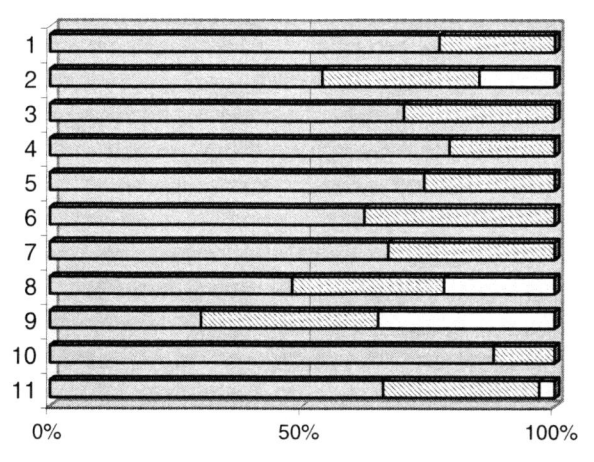

Fig. 4.33 "Solitary parts" A_s (*solid*) and $H_s/2$ (*hatched*) of the freak wave amplitudes for 10 records from the North Alwyn platform and the New Year Wave

ously proves the important role of the nonlinear modulation effect in the freak-wave occurrence.

The application of the IST method to the analysis of water wave groups may be improved by use of the Creamer et al. (1989) transform that takes into account the nonlinear bound corrections that are not described by the NLS envelope equation. Higher-order integrable (or nearly integrable) versions of the envelope equation may be employed to describe more accurately the envelope solitary solutions (Schober and Calini 2008).

In the present analysis, we have employed the window Fourier transform to determine the wave frequency. Wavelets provide an alternative improved way to estimate the mean wave scale. They have been used by various authors: among them we cite Mori et al. (2002), Chien et al. (2002), Paprota et al. (2003), Scott et al. (2005). These studies present different occurrences of rogue waves in wavelet planes. Chien et al. (2002) distinguish freak waves generated by wind waves (unimodal spectrum with strong grouping phenomenon) and bimodal waves caused by interaction of two wave systems (say, wind waves and swell). There also exist a large amount of multimodal waves that have many energetic areas in the wavelet spectrum. Although the wavelet analysis may catch the transient change of wave parameters better than the Fourier transform, the wavelet spectra are more difficult to interpret. The shapes of the prototype functions ("mother wavelets") are often very similar (or identical) to the NLS envelope soliton, hence the application of the IST analysis in combination with the exact theory may prove to be very efficient.

The three-dimensionality of rogue waves can help to identify their origin, as is discussed in Sect. 4.6. Although the development of air-, ship- and satellite-borne SAR measurements and the associated methods of analysis are very promising (see Rosenthal 2005), until now there have been very few results concerning 3D observations. It needs further improvement and justification to enable regular measurements and analysis (see Dysthe et al. 2008).

4.7.2 Statistics from Registrations of Natural Rogue Waves

Most available long-run instrumental registrations are made over relatively deep-water areas (see Fig. 1.3). Although the number of measured rogue waves is in the hundreds, these waves are measured under very different conditions and obviously do not satisfy the stationary random process requirement. That is why the results of their statistical analysis are often dubious. This doubt is indirectly confirmed by conflicting conclusions of different investigations about the probability of highest waves registered by gauges. Rogue waves are found to occur much more frequently than is foreseen by the Rayleigh distribution function in studies (Mori 2004, Pinho et al. 2004, Stansell 2004, Liu and MacHutchon 2006). This distribution, however, fits natural data reported quite well in Mori et al. (2002). The freak wave phenomenon is rarer than it follows from the Rayleigh distribution function according to Chien et al. 2002, Paprota et al. 2003, both for relatively shallow water). Stansell (2004) has undertaken a careful analysis of the records from the viewpoint of statistical stationarity, and reported on about 300 times more frequent occurrence of the highest measured wave ($AI = 3.19$) than it could be expected from the Reyleigh statistics. Similar estimates may be found in the paper by Mori and Janssen (2006a). Although some theoretical relations are suggested by the authors to describe the results, the general disagreement between the results about the rogue-wave probability obviously makes the conclusions about the quantitative rogue-wave probability estimation premature. Thus, the present database of rogue waves cannot answer the question about the true probability of rogue waves. The more or less accepted opinion about the statistical description of observed extreme waves is as follows: the high-order statistical models in general are able to describe many huge waves, although a population exists of "true rogue waves" that do not satisfy the classical statistical description.

It has already been discussed that the scientific community tries to fill in the lack of in situ data by numerical data obtained from computational runs. To do this, it is necessary to ensure that the dynamics described by the computer models are similar to real ocean dynamics. The main result achieved through the numerical simulations of irregular surface waves consists of an increase of the rogue wave probability when the Benjamin-Feir index grows. Hence, this parameter has been considered as a possible good indicator of high probability of freak-wave occurrence. Therefore, the first question that should be answered is: does the probability of extreme sea-waves exhibit a dependence on the BFI? The answer is actually not straightforward. The BFI seems to be a promising parameter for evaluating the danger of extreme sea waves. As it was demonstrated by numerical simulation, a strong correlation exists between high wave probability and BFI. Nevertheless, its practical use seems to be still not fully operational. The BFI is a complex parameter, roughly speaking, reflecting the typical wave height (or corresponding dimensionless parameter, "steepness") and spectral bandwidth (or number of waves in a group, which is the inverse value) (see Sect. 4.4). Surprisingly, it is found that the probability of occurrence of freak waves is only weakly dependent on the significant wave height, significant wave steepness, and spectral bandwidth (Stansell 2004, Olagnon and Prevosto 2005, Olagnon and Prevosto 2005). Furthermore, Melville et al. (2005)

remark that the threshold in the abnormality index, AI, does not correspond to equivalent thresholds in either the skewness or excess kurtosis.

In order to confirm the adequacy of a selected parameter to be used in warnings of risk occurrence, it is necessary to ensure that it is sensitive to the presence of rogue waves, and that it can be robustly computed. Olagnon and Magnusson (2004, 2005) note that the BF indices (defined in Olagnon and Magnusson 2004 in different ways) and the peakedness factor of the JONSWAP spectrum exhibit particularly poor robustness. High natural variability of the BF index might be a consequence of the difficulty to obtain stable estimators when considering short in situ records.

The investigation of the robustness of some popular statistical parameters (wave height, crest height, period, steepness, kurtosis, BFI, parameters of the spectral shape) performed by Olagnon and Magnusson (2004) reports that only the kurtosis exhibits a sufficient correlation with normalized crest height to allow considering it as a parameter to be monitored. However, the kurtosis is directly influenced by the presence of extreme waves. We should emphasize here that from the theoretical point of view the kurtosis and the BFI are related through Eq. (4.108).

From a practical viewpoint, a parameter must vary on a characteristic time scale significantly larger than the wave period. Otherwise, the variation of the parameter will merely be a detector of the rogue wave and cannot be used for forecasting. Olagnon and Prevosto (2005) report that the change of the kurtosis value registered at the instant of a high wave occurrence can be satisfactorily explained by the high wave alone, and that no further relationship can be found at larger time scales. If the maximum wave is removed from the kurtosis computation, and kurtosis is estimated from the remaining of the record, no further correlation between the kurtosis and the maximum wave height can be seen. Therefore, Olagnon and Prevosto (2005) conclude that the Benjamin-Feir instability is very local and is not reflected by statistics at the time scale of a sea state. The deviations that they could observe for some spectral parameters close to occurrences of extreme waves were well within the natural range of variability. They could not identify any special feature on the time-histories of the BFI that might have some chance of being related to rogue wave occurrence.

A possible explanation of the discrepancy between numerical studies and natural observations may be due to the typically unidirectional wave propagation (long-crested waves) studied in the majority of the numerical computations, while the natural sea waves are essentially short-crested. The evidence of two qualitatively different sea wave regimes (long- and short-crested) that result in very different statistics is formulated in recent papers (Haver 2005, Gramstad and Trulsen 2007, Dysthe et al. 2008) (see Sect. 4.6), and is becoming supported by theoretical studies and numerical simulations as well. These studies will obviously guide the focus of future research.

List of Notations

A	amplification factor
A	wave amplitude
A_{br}	amplitude of the breather

A_{es}	amplitude of the envelope soliton
A_{pw}	amplitude of the plane wave
A_{wp}	dimensional amplitude of the wave packet
AI	abnormality index
C_{gr}	group velocity
C_{LW}	long wave velocity
C_{ph}	phase velocity
d	dimensionless water depth
$\tilde{d}$	depth parameter
D	water depth
D/DT	material derivative
g	acceleration due to gravity
H	wave height
H_s	significant wave height
I_{BF}	Benjamin-Feir index
$\mathbf{k} = (p,q)$	dimensionless wave vector
$\mathbf{K} = (K_X, K_Y)$	wave vector
K	wavenumber
M	mass integral
M	order of perturbation series in the HOSM approach
$\mathbf{n}$	unit vector normal to the boundary
n_t, n_x	number of individual wave in the time series or wave snapshot
N_s	soliton number
p	dimensionless pressure
p_a	dimensionless atmosphere pressure
P	pressure
P_a	atmosphere pressure
$q(x,t)$	dimensionless envelope amplitude in the NLS equation
s	wave steepness
t	dimensionless time
T	time
T_{br}	period of the breather
T_f	focusing time
U_w	wind velocity
V_{br}	velocity of the breather
V_{es}	velocity of the envelope soliton
V_{wp}	dimensional velocity of the wave packet
(x, y, z)	dimensionless coordinates
(X, Y, Z)	coordinates
X_f	focusing length
$\phi(X, Y, Z, T)$	velocity potential
$\eta(X, Y, T)$	surface elevation
$\varphi(x, y, z, t)$	dimensionless velocity potential
κ	kurtosis
λ	eigenvalue of the associated scattering problem
ρ_a	atmosphere density

σ	growth rate
σ	standard deviation, σ^2 is the variance
Ω	cyclic wave frequency
Ω_p	mean wave frequency
$\zeta(x, y, t)$	dimensionless surface displacement
$\partial\Omega_{FS}$	free surface
$\partial\Omega_{SB}$	solid boundaries
∇	gradient operator

List of Acronyms

BF	Benjamin-Feir
BFI	Benjamin-Feir Index
BIEM	Boundary Integral Equation Method
DS	Davey-Stewartson system
HOSM	High Order Spectral Method
NLS	Nonlinear Schrödinger equation
SWE	Steep Wave Event

References

Ablowitz MJ, Hammack J, Henderson D, Schober CM (2000) Modulated periodic Stokes waves in deep water. Phys Rev Lett 84:887–890

Ablowitz MJ, Hammack J, Henderson D, Schober CM (2001) Long-time dynamics of the modulational instability of deep water waves. Phys D 152–153:416–433

Ablowitz MJ, Herbst BM (1990) On homoclinic structure and numerically induced chaos for the Nonlinear Schrödinger equation. SIAM J Appl Math 50:339–351

Ablowitz MJ, Kaup DJ, Newell AC, Segur H (1974) The inverse scattering transform – Fourier analysis for nonlinear problems. Stud Appl Math 53:249–315

Ablowitz MJ, Segur H (1979) On the evolution of packets of water waves. J Fluid Mech 92:691–715

Akhmediev NN, Ankiewicz A (1997) Solitons. Nonlinear pulses and beams. Chapman & Hall, Florida

Akhmediev NN, Eleonskii VM, Kulagin NE (1985) Generation of periodic trains of picosecond pulses in an optical fiber: exact solutions. Sov Phys J Exp Theor Phys 62:894–899

Akhmediev NN, Eleonskii VM, Kulagin NE (1987) Exact first-order solutions of the nonlinear Schrödinger equation. Theor Math Phys (USSR) 72:809–818

Alber IE (1978) The effects of randomness on the stability of two-dimensional surface wavetrain. Proc Roy Soc Lond A 363:525–546

Alber IE, Saffman PG (1978) Stability of random nonlinear water waves with finite bandwidth spectra. TRW Def and Space Syst Group Rep No 31326-6035-RU-00bv

Annenkov SYu, Badulin SI (2001) Multi-wave resonances and formation of high-amplitude waves in the ocean. In: Olagnon M, Athanassoulis GA (eds) Rogue Waves 2000, Ifremer, France, pp 205–213

Badulin SI, Shrira VI, Kharif C, Ioualalen M (1995) On two approaches to the problem of instability of short-crested water waves. J Fluid Mech 303:297–326

Baldock TE, Swan C, Taylor PH (1996) A laboratory study of nonlinear surface waves on water. Phil Trans Roy Soc Lond A 354:649–676

Banner ML, Song J (2002) On determining the onset and strength of breaking for deep water waves. Part II: Influence of wind forcing and surface shear. J Phys Oceanogr 32:2559–2570

Banner ML, Tian X (1998) On the determination of the onset of breaking for modulating surface gravity water waves. J Fluid Mech 367:107–137

Bateman WJD, Swan C, Taylor PH (2001) On the efficient numerical simulation of directionally spread surface water waves. J Comput Phys 174:277–305

Benjamin TB, Feir JE (1967) The desintegration of wave trains on deep water. Part 1. Theory. J Fluid Mech 27:417–430

Benney DJ, Roskes GJ (1969) Wave instabilities. Stud Appl Math 48:377–385

Bliven LF, Huang NE, Long SR (1986) Experimental study of the influence of wind on Benjamin-Feir sideband instability. J Fluid Mech 162:237–260

Bonnefoy F, de Reilhac PR, Le Touzé D, Ferrant P (2005) Numerical and physical experiments of wave focusing in short-crested seas. In: Proc. 14th Aha Huliko'a Winter Workshop, Honolulu, Hawaii, 2005

Burzlaff J (1988) The soliton number of optical soliton bound states for two special families of input pulses. J Phys A: Math Gen 21:561–566

Calini A, Schober CM (2002) Homoclinic chaos increases the likelihood of rogue wave formation. Phys Lett A 298:335–349

Caponi EA, Saffman PG, Yuen HC (1982) Instability and confined chaos in a nonlinear dispersive wave system. Phys Fluids 25:2159–2166

Chien H, Kao C-C, Chuang LZH (2002) On the characteristics of observed coastal freak waves. Coast Eng J 44:301–319

Clamond D, Francius M, Grue J, Kharif C (2006) Long time interaction of envelope solitons and freak wave formations. Eur J Mech B/Fluids 25:536–553

Clamond D, Grue J (2001) A fast method for fully nonlinear water-wave computations. J Fluid Mech 447:337–355

Clamond D, Grue J (2002) Interaction between envelope solitons as a model for freak wave formations. Pt. 1: Long time interaction. C R Mecanique 330:575–580

Clarke S, Grimshaw R, Miller P, Pelinovsky E, Talipova T (2000) On the generation of solitons and breathers in the modified Korteweg – de Vries equation. Chaos 10:383–392

Crawford DR, Saffman PG, Yuen HC (1980) Evolution of a random inhomogeneous field of nonlinear deep-water gravity waves. Wave Motion 2:1–16

Creamer DB, Henyey F, Schult R, Wright J (1989) Improved linear representation of ocean surface waves. J Fluid Mech 205:135–161

Davey A, Stewartson K (1974) On the three-dimensional packets of surface waves. Proc Roy Soc Lond A 338:101–110

Desaix M, Anderson D, Lisak M, Quiroga-Teixeiro ML (1996) Variationally obtained approximate eigenvalues of the Zakharov-Shabat scattering problem for real potentials. Phys Lett A 212:332–338

Dhar AK, Das KP (1991) Fourth-order nonlinear evolution equation for two Stokes wave trains in deep water. Phys Fluids A 3(12):3021–3026

Dias F, Kharif C (1999) Nonlinear gravity and capillary-gravity waves. Annu Rev Fluid Mech 31:301–346

Dold JW (1992) An efficient surface-integral algorithm applied to unsteady gravity waves. J Comput Phys 193:90–115

Dold JW, Peregrine DH, (1986) Water wave modulation. In: Proc 20th Int Conf Coast Eng, ASCE, Taipei 1:163–175

Dommermuth D, Yue DKP (1987) A high-order spectral method for the study of nonlinear gravity waves. J Fluid Mech 184:267–288

Drazin PG, Johnson RS (1989) Solitons: an Introduction. Cambridge University Press, Cambridge

Ducrozet G, Bonnefoy F, Le Touzé D, Ferrant P (2007) 3-D HOS simulations of extreme waves in open seas. Nat Hazards Earth Syst Sci 7:109–122

References

Duin CA van (1999) The effect of non-uniformity of modulated wavepackets on the mechanism of Benjamin – Feir instability. J Fluid Mech 399:237–249

Dyachenko AI, Zakharov VE (2005) Modulation instability of stokes wave –> freak wave. J Exp Theor Phys Lett 81:255–259

Dysthe K, Krogstad HE, Müller P (2008) Oceanic rogue waves. Annu Rev Fluid Mech 40:287–310

Dysthe KB (1979) Note on a modification to the nonlinear Schrödinger equation for application to deep water waves. Proc Roy Soc London A 369:105–114

Dysthe KB, Trulsen K (1999) Note on breather type solutions of the NLS as a model for freak-waves. Physica Scripta T82:48–52

Dysthe KB, Trulsen K, Krogstad HE, Socquet-Juglard H (2003) Evolution of a narrow-band spectrum of random surface gravity waves. J Fluid Mech 478:1–10

Engelbrecht JK, Fridman VE, Pelinovski EN (1988) Nonlinear evolution equations. Jeffrey A (ed). Longman, London

Fochesato C, Grilli S, Dias F (2007) Numerical modeling of extreme rogue waves generated by directional energy focusing. Wave Motion 44:395–416

Francius M, Kharif C (2006) Three-dimensional instabilities of periodic gravity waves in shallow water. J Fluid Mech 561:417–437

Fructus D, Kharif C, Francius M, Kristiansen Ø, Clamond D, Grue J (2005) Dynamics of crescent water wave patterns. J Fluid Mech 537:155–186

Gibson RS, Swan C, Tromans PS (2007) Fully nonlinear statistics of wave crest elevation calculated using a spectral response surface method: Application to unidirectional sea states. J Phys Oceanogr 37:3–15

Gramstad O, Trulsen K (2007) Influence of crest and group length on the occurrence of freak waves. J Fluid Mech 582:463–472

Grue J, Jensen A (2006) Experimental velocities and accelerations in very steep wave events in deep water. Eur J Mech B/Fluids 25:554–564

Hasimoto H, Ono H (1972) Nonlinear modulation of gravity waves. J Phys Soc Japan 33:805–811

Haver S (2005) Freak waves: a suggested definition and possible consequences for marine structures. In: Olagnon M, Prevosto M (eds) Rogue Waves 2004, Ifremer, France

Henderson KL, Peregrine DH, Dold JW (1999) Unsteady water wave modulations: Fully nonlinear solutions and comparison with the nonlinear Schrödinger equation. Wave Motion 29:341–361

Ioualalen M, Kharif C (1994) On the subharmonic instabilities of steady three-dimensional deep water waves. J Fluid Mech 262:265–291

Islas AL, Schober CM (2005) Predicting rogue waves in random oceanic sea states. Phys Fluids 17:031701-1–4

Janssen PAEM (2003) Nonlinear four-wave interactions and freak waves. J Phys Oceanogr 33:863–884

Jeffreys H (1925) On the formation of wave by wind. Proc Roy Soc A 107:189–206

Jensen A, Sveen JK, Grue J, Richon JB, Gray C (2001) Accelerations in water waves by extended particle image velocimetry. Exp in Fluids 30:500–510

Johannessen TB, Swan C (2001) A laboratory study of the focusing of transient and directionally spread surface water waves. Proc Roy Soc Lond A 457:971–1006

Johnson RS (1977) On the modulation of water waves in the neighbourhood of $kh \approx 1.363$. Proc Roy Soc Lond A 357:131–141

Johnson RS (1997) A modern introduction to the mathematical theory of water waves. Cambridge Univ Press

Kakutani T, Michihiro K (1983) Marginal state of modulational instability – Note on Benjamin – Feir instability. J Phys Soc Japan 52:4129–4137

Kaup DJ, Malomed BA (1995) Variational principle for the Zakharov-Shabat equations. Phys D 84:319–328

Kawata T, Inoue H (1978) Inverse scattering method for the nonlinear evolution equations under nonvanishing conditions. J Phys Soc Japan 44:1722–1729

Kharif C, Ramamonjiarisoa A (1988) Deep water gravity waves instabilities at large steepness. Phys Fluids 31:1286–1288

Kharif C, Pelinovsky E, Talipova T, Slunyaev A (2001) Focusing of nonlinear wave groups in deep water. J Exp Theor Phys Lett 73:170–175

Kharif C, Pelinovsky E (2006) Freak waves phenomenon: physical mechanisms and modelling. In: Grue J, Trulsen K (eds) Waves in geophysical fluids: Tsunamis, Rogue waves, Internal waves and Internal tides. CISM Courses and Lectures No. 489. Springer Wein, New York

Kharif C, Giovanangeli JP, Touboul J et al (2008) Influence of wind on extreme wave events: experimental and numerical approaches. J Fluid Mech 594:209–247

Kimmoun O, Branger H, Kharif C (1999) On short-crested waves: experimental and analytical investigations. Eur J Mech B/Fluids 18:889–930

Krasitskii VP (1994) On reduced equations in the Hamiltonian theory of weakly nonlinear surface waves. J Fluid Mech 272:1–30

Krein MG (1955) Foundations of the theory of λ-zones of stability of a canonical system of linear differential equations with periodic coefficients. Am Math Soc Trans Ser 2 120:1–70

Kristiansen Ø, Fructus D, Clamond D, Grue J (2005) Simulations of crescent water wave patterns on finite depth. Phys Fluids 17:064101-1–15

Kusuba T, Mitsuyasu M (1986) Nonlinear instability and evolution of steep water waves under wind action. Rep Res Inst Appl Mech Kyushu Univ 33:33–64

Kuznetsov EA (1977) To the question of solitons in parametrically unstable plasma. Dokl USSR 236:575–577. (In Russian)

Landau LD, Lifshitz EM (1980) Course of theoretical physics. Volume 3: Quantum mechanics (non-relativistic theory). Pergamon Press Ltd, Hungary

Li JC, Hui WH, Donelan MA (1987) Effects of velocity shear on the stability of surface deep water wave trains. In: Nonlinear Water Waves (IUTAM Symp). Springer Verlag, Heidelberg, 213–220

Lighthill MJ (1965) Contributions to the theory of waves in nonlinear dispersive systems. J Inst Math Appl 1:269–306

Liu PC, MacHutchon KR (2006) Are there different kinds of rogue waves? In Proc 25th Int Conf OMAE 2006, Hamburg, Germany, 2006, OMAE2006-92619:1-6

Longuet-Higgins M (1978a) The instabilities of gravity waves of finite amplitude in deep water. I. Superharmonics. Proc Roy Soc Lond Ser A 360:471–488

Longuet-Higgins M (1978b) The instabilities of gravity waves of finite amplitude in deep water. II. Subharmonics. Proc Roy Soc Lond Ser A 360:489–505

Longuet-Higgins MS (1985) Bifurcation in gravity waves. J Fluid Mech 151:457–475

Longuet-Higgins MS, Stewart RW (1960) Changes in the form of short gravity waves on long waves and tidal currents. J Fluid Mech 8:565–583

Ma Y-Ch (1979) The perturbed plane-wave solutions of the cubic Schrödinger equation. Stud Appl Math 60:43–58

MacKay RS, Saffman PG (1986) Stability of water waves. Proc Roy Soc Lond A 406:115–125

Massel SR (1996) Ocean surface waves: their physics and prediction. World Scientific Publishing Co Pte Ltd, Singapore

McLean JW (1982a) Instabilities of finite amplitude water waves. J Fluid Mech 114:315–330

McLean JW (1982b) Instabilities of finite-amplitude gravity waves on water of finite depth. J Fluid Mech 114:331–341

McLean JW, Ma YC, Martin DU et al (1981) Three-dimensional instability of finite-amplitude water waves. Phys Rev Lett 46:817–820

Melville WK, Romero L, Kleiss JM (2005) Extreme wave events in the Gulf of Tehuantepec. In: Proc. 14th Aha Huliko'a Winter Workshop, Honolulu, Hawaii, 2005

Melville WK (1982) The instability and breaking of deep-water waves. J Fluid Mech 115:165–185

Mori N, Liu PC, Yasuda T (2002) Analysis of freak wave measurements in the Sea of Japan. Ocean Eng 29:1399–1414

Mori N (2004) Occurrence probability of a freak wave in a nonlinear wave field. Ocean Eng 31:165–175

Mori N, Janssen PAEM (2006a) Freak wave prediction from directional spectra. In: Proc 30th Int Conf Coast Eng, ASCE, pp 714–725

Mori N, Janssen PAEM (2006b) On kurtosis and occurrence probability of freak waves. J Phys Oceanogr 36:1471–1483

Nakamura A, Hirota R (1985) A new example of explode-decay solitary waves in one dimension. J Phys Soc Japan 54:491–499

Newell AC (1981) Solitons in mathematics and physics. Univ Arizona: Soc Ind Appl Math

Novikov S, Manakov SV, Pitaevskii LP, Zakharov VE (1984) Theory of Solitons: the Inverse Scattering Method. Consult Bureau, New York

Olagnon M, Magnusson AK (2004) Sensitivity study of sea state parameters in correlation to extreme wave occurrences. In: Proc. 14th Int Offshore and Polar Eng Conf ISOPE, Toulon, France, 2004, pp 18–25

Olagnon M, Magnusson AK (2005) Spectral parameters to characterize the risk of rogue waves occurrence in a sea state. In: Olagnon M, Prevosto M (eds) Rogue Waves 2004, Ifremer, France

Olagnon M, Prevosto M (2005) Are rogue waves beyond conventional predictions? In: Olagnon M, Prevosto M (eds) Rogue Waves 2004, Ifremer, France

Onorato M, Osborne AR, Serio M (2002) Extreme wave events in directional, random oceanic sea states. Phys Fluids 14:L25–28

Onorato M, Osborne AR, Serio M, Bertone S (2001) Freak waves in random oceanic sea states. Phys Rev Lett 86:5831–5834

Onorato M, Osborne AR, Serio M, Cavaleri L (2005) Modulational instability and non-Gaussian statistics in experimental random water-wave trains. Phys Fluids 17:078101-1–4

Onorato M, Osborne AR, Serio M et al (2004) Observation of strongly non-Gaussian statistics for random sea surface gravity waves in wave flume experiments. Phys Rev E 70:067302-1–4

Onorato M, Osborne AR, Serio M (2006a) Modulational instability in crossing sea states: A possible mechanism for the formation of freak waves. Phys Rev Lett 96: 014503-1–4

Onorato M, Osborne AR, Serio M et al (2006b) Extreme waves, modulational instability and second order theory: wave flume experiments on irregular waves. Eur J Mech B / Fluids 25:586–601

Osborne AR, Onorato M, Serio M (2000) The nonlinear dynamics of rogue waves and holes in deep water gravity wave trains. Phys Lett A 275:386–393

Osborne AR, Onorato M, Serio M (2005) Nonlinear Fourier analysis of deep-water, random surface waves: Theoretical formulation and experimental observations of rogue waves. In: Proc 14th Aha Huliko'a Winter Workshop, Honolulu, Hawaii, 2005

Osborne AR, Petti M (1994) Laboratory-generated, shallow-water surface waves: Analysis using the periodic, inverse scattering transform. Phys Fluids 6:1727–1744

Paprota M, Przewłócki J, Sulisz W, Swerpel BE (2003) Extreme waves and wave events in the Baltic Sea. In: Rogue Waves: Forecast and Impact on Marine Structures. GKSS Research Center, Geesthacht, Germany

Pelinovsky EN, Slunyaev AV, Talipova TG, Kharif C (2003) Nonlinear parabolic equation and extreme waves on the sea surface. Radiophysics and Quantum Electronics 46:451–463

Peregrine DH (1983) Water waves, nonlinear Schrödinger equations and their solutions. J Austral Math Soc Ser B 25:16–43

Pinho de UF, Liu PC, Ribeiro CEP (2004) Freak waves at Campos Basin, Brazil. Geofizika 21:53–67

Rapp RJ, Melville WK (1990) Laboratory measurements of deep water breaking waves. Phil Trans Roy Soc Lond A 331:735–800

Rosenthal W (2005) Results of the MAXWAVE project. In: Proc. 14th Aha Huliko'a Winter Workshop, Honolulu, Hawaii, 2005. http://www.soest.hawaii.edu/PubServices/2005pdfs/Rosenthal.pdf. Accessed 14 March 2008

Roskes GJ (1976) Some nonlinear multiphase interactions. Stud Appl Math 55:231–238

Ruban VP (2007) Nonlinear stage of the Benjamin–Feir instability: three-dimensional coherent structures and rogue waves. Phys Rev Lett 99:044502-1–4

Satsuma J, Yajima N (1974) Initial value problems of one-dimensional self-modulation of nonlinear waves in dispersive media. Suppl Progr Theor Phys 55:284–306

Schober CM, Calini A (2008) Extreme Ocean Waves Rogue waves in higher order nonlinear Schrödinger models. In: Pelinovsky E, Kharif C (eds) Springer

Scott N, Hara T, Hwang PA, Walsh EJ (2005) Directionality and crest length statistics of steep waves in open ocean waters. J Atmos Ocean Tech 22:272–281

Sedletsky Yu (2006) A new type of modulation instability of Stokes waves in the framework of an extended NSE system with mean flow. J Phys A: Math Gen 39:L529–L537

Sedletsky YuV (2003) The fourth-order nonlinear Schrödinger equation for the envelope of Stokes waves on the surface of a finite-depth fluid. J Exp Theor Phys 97:180–193

Segur H, Henderson D, Carter J et al (2005) Stabilizing the Benjamin – Feir instability. J Fluid Mech 539:229–271

Shukla PK, Kourakis I, Eliasson B et al (2006) Instability and evolution of nonlinearly interacting water waves. Phys Rev Lett 97:094501-1-4

Skandrani C, Kharif C, Poitevin J (1996) Nonlinear evolution of water surface waves: The frequency down-shift phenomenon. Contemp Math 200:157–171

Slunyaev A, Kharif C, Pelinovsky E, Talipova T (2002) Nonlinear wave focusing on water of finite depth. Phys D 173:77–96

Slunyaev A (2006) Nonlinear analysis and simulations of measured freak wave time series. Eur J Mech B / Fluids 25:621–635

Slunyaev A, Pelinovsky E, Guedes Soares C (2005) Modeling freak waves from the North Sea. Appl Ocean Res 27:12–22

Slunyaev AV (2001) The initial problem for the modified Korteweg – de Vries equation on a pedestal: birth of solitons and breathers. Proc Ac Eng Sc Russ Fed 2:166–175. (In Russian)

Slunyaev AV (2005) A high-order nonlinear envelope equation for gravity waves in finite-depth water. J Exp Theor Phys 101:926–941

Socquet-Juglard H, Dysthe KB, Trulsen K, Krogstad HE, Liu J (2005) Probability distributions of surface gravity waves during spectral changes. J Fluid Mech 542:195–216

Stansell P (2004) Distributions of freak wave heights measured in the North Sea. Appl Ocean Res 26:35–48

Stiassnie M (1984) Note on the modified nonlinear Schrödinger equation for deep water waves. Wave Motion 6:431–433

Stiassnie M, Shemer L (1984) On medications of the Zakharov equation for surface gravity waves. J Fluid Mech 143:47–67

Su MY (1982) Three-dimensional deep-water waves. Part 1. Experimental measurement of skew and symmetric wave patterns. J Fluid Mech 124:73–108

Su MY, Bergin M, Marler P, Myrick R (1982) Experiments on nonlinear instabilities and evolution of steep gravity-wave trains. J Fluid Mech 124:45–72

Tajiri M, Watanabe Y (1998) Breather solutions to the focusing nonlinear Schrödinger equation, Phys Rev E 57:3510–3519

Tanaka M (2007) Numerical study on the occurrence probability of freak waves in two- and three-dimensional quasi-stationary wave fields. Electronic Preprint.

Tayfun M (1980) Narrow–band nonlinear sea waves. J Geophys Res 85 C3:1548–1552

Toffoli A, Lefèvre JM, Monbaliu J, Bitner-Gregersen E (2004) Dangerous sea-states for marine operations. In: Proc. 14th Int Offshore and Polar Eng Conf ISOPE, Toulon, France, 2004, pp 85–92

Touboul J, Giovanangeli JP, Kharif C, Pelinovsky E (2006) Freak waves under the action of wind: Experiments and simulations. Eur J Mech B/Fluids 25:662–676

Touboul J, Pelinovsky E, Kharif C (2007) Nonlinear focusing wave groups on current. J Korean Soc Coastal and Ocean Eng 9:222–227

Trulsen K (2006) Weakly nonlinear and stochastic properties of ocean wave fields: application to an extreme wave event. In: Grue J, Trulsen K (eds) Waves in geophysical fluids: Tsunamis, Rogue waves, Internal waves and Internal tides. CISM Courses and Lectures No. 489. Springer Wein, New York

Trulsen K, Dysthe KB (1996) A modified nonlinear Schrödinger equation for broader bandwidth gravity waves on deep water. Wave Motion 24:281–289

Trulsen K, Kliakhandler I, Dysthe KB, Velarde MG (2000) On weakly nonlinear modulation of waves on deep water. Phys Fluids 12:2432–2437

References

Voronovich VV, Shrira VI, Thomas G (2008) Can bottom friction suppress 'freak wave' formation? J Fluid Mech 604:263–296

Waseda T, Tulin MP (1999) Experimental study of the stability of deepwater wave trains including wind effects. J Fluid Mech 401:55–84

West BJ, Brueckner KA, Janda RS et al (1987) A new numerical method for surface hydrodynamics. J Geophys Res 92:11803–11824

Whitham GB (1967) Variational methods and applications to water waves. Proc Roy Soc Lond A 229:6–25

Wu CH, Yao A (2004) Laboratory measurements of limiting freak waves on currents. J Geophys Res 109:C12002-1–18

Yasuda T, Mori N (1997) Roles of sideband instability and mode coupling in forming a water-wave chaos. Wave Motion 26:163–185

Zabusky NJ, Kruskal MD (1965) Interaction of solutions in a collisionless plasma and recurrence of initial states. Phys Rev Lett 15:240–243

Zakharov V (1968) Stability of periodic waves of finite amplitude on a surface of deep fluid. J Appl Mech Tech Phys 2:190–194

Zakharov VE, Dias F, Pushkarov AN (2006a) One-dimensional wave turbulence. Phys Rep 398:1–65

Zakharov VE, Dyachenko AI, Prokofiev AO (2006b) Freak waves as nonlinear stage of Stokes wave modulation instability. Eur J Mech B / Fluids 25:677–692

Zakharov VE, Shabat AB (1972) Exact theory of two-dimensional self-focussing and one-dimensional self-modulation of waves in nonlinear media. Sov Phys J Exp Theor Phys 34:62–69

Chapter 5
Shallow-Water Rogue Waves

When the sea becomes shallow, the water flow induced by surface waves is almost uniform with depth. Thus, properties of shallow water waves are radically different from those in deep water: the wave dispersion is weak, and waves now "feel" the seafloor. Nonlinearity leads to strong correlation between spectral components supporting existence of various wave shapes such as solitons, cnoidal waves, and undular bores. The interaction of water waves with variable bathymetry and coastal lines modifies the wave regime in shallow water and influences rogue wave formation. This chapter is devoted to the description of theoretical models of shallow-water freak waves.

5.1 Nonlinear Models of Shallow-Water Waves

The basic 3D hydrodynamic models are effective for studying wave processes in relatively small basins due to limited computer resources. This is why various depth-averaged models (2D) are popular to describe wind wave processes in the coastal zone of seas and oceans, and sometimes for transoceanic propagation of large-scale waves (such as a tsunami). A straightforward way to derive nonlinear dispersive models of shallow-water waves is to use the Euler equation written for potential flow

$$\mathbf{U} = \nabla \phi, \quad W = \partial \phi / \partial Z. \tag{5.1}$$

All vector operations hereafter act in the horizontal plane, so that $\nabla = (\partial/\partial X, \partial/\partial Y)^t$, $\Delta = \nabla \cdot \nabla$ and $\mathbf{U} = (U, V)$ (see geometry in Fig 2.1). Then, the Laplace equation (2.13) has the form

$$\Delta \phi + \frac{\partial^2 \phi}{\partial Z^2} = 0, \tag{5.2}$$

with the boundary condition (2.31) at the uneven bottom, $Z = -D(X,Y)$,

$$\partial \phi / \partial Z + \nabla \phi \cdot \nabla D = 0, \tag{5.3}$$

and kinematic and dynamic conditions on the free surface, $Z = \eta(X,Y,T)$ (see Chap. 2),

$$\frac{\partial \phi}{\partial Z} = \frac{d\eta}{dT} = \frac{\partial \eta}{\partial T} + \nabla \phi \cdot \nabla \eta \text{ on } Z = \eta, \qquad (5.4)$$

$$\frac{\partial \phi}{\partial T} + \frac{1}{2}\left(\frac{\partial \phi}{\partial Z}\right)^2 + g\eta = 0 \text{ on } Z = \eta. \qquad (5.5)$$

Since potential flow is governed by a harmonic function, it can be differentiated with respect to all its arguments and expanded as a Taylor series with respect to the vertical coordinate centered at $Z = -D$,

$$\phi(X,Y,Z,T) = \sum_{n=0}^{\infty} q_n(X,Y,T)(Z+D)^n. \qquad (5.6)$$

Substitution of Eq. (5.6) into the Laplace equation (5.2) yields the recurrence correlations for the unknown functions q_n,

$$(n+2)(n+1)q_{n+2} + \Delta q_n + 2(n+1)\nabla q_{n+1}\nabla D \\ + (n+1)q_{n+1}\Delta D + (n+2)(n+1)q_{n+2}(\nabla D)^2 = 0, \qquad (5.7)$$

so that only two of them (namely, q_0 and q_1) are independent. Specifically, q_2 is given by

$$q_2 = -\frac{\Delta q_0 + 2\nabla q_1 \nabla D + q_1 \Delta D}{2[1+(\nabla D)^2]}. \qquad (5.8)$$

By substituting series (5.6) into the boundary condition on the bottom (5.3) and using Eq. (5.7), we may deduce the following relation between q_1 and q_0

$$q_1 = -\frac{\nabla q_0 \nabla D}{[1+(\nabla D)^2]}. \qquad (5.9)$$

Thus, the series (5.6) is completely determined by only one function, $q_0(X,Y,T)$. Boundary conditions on the free surface (5.4) and (5.5) provide equations for η and ∇q_0. The physical meaning of ∇q_0 is the bottom velocity (for a flat floor). When the depth-averaged velocity

$$\mathbf{u}(X,Y,T) = \frac{1}{D+\eta}\int_{-D}^{\eta} \nabla\phi(X,Y,Z,T)dZ \qquad (5.10)$$

is chosen as the "physical" horizontal velocity, then the value of ∇q_0 can be calculated from (5.6) approximately as

$$\nabla q_0 = \mathbf{u} + \frac{D+\eta}{2}\mathbf{u}\Delta D + (D+\eta)(\nabla D\nabla)\mathbf{u} + (\nabla D)^2\mathbf{u} + \frac{(D+\eta)^2}{6}\Delta\mathbf{u} + \ldots, \qquad (5.11)$$

where the iteration procedure employs a small parameter D/λ, where λ is the wavelength characterizing the "shallowness" of long-water waves. After substitution of series (5.11), Eqs. (5.4) and (5.5) result in equations for the fully nonlinear

5.1 Nonlinear Models of Shallow-Water Waves

weakly dispersive theory (see details in Green and Naghdi 1976, and Zheleznyak and Pelinovsky 1985)

$$\frac{\partial \eta}{\partial T} + \nabla \cdot [(D+\eta)\mathbf{u}] = 0, \tag{5.12}$$

$$\frac{\partial \mathbf{u}}{\partial T} + (\mathbf{u}\nabla)\mathbf{u} + g\nabla \eta = \mathbf{F}, \tag{5.13}$$

where $\mathbf{F}$ characterizes weak dispersion

$$\mathbf{F} = \frac{1}{D+\eta} \nabla \left[\frac{(D+\eta)^3}{3} R + \frac{(D+\eta)^2}{2} Q \right] - \nabla D \left[\frac{D+\eta}{2} R + Q \right], \tag{5.14}$$

$$R = \frac{\partial}{\partial T} \nabla \cdot \mathbf{u} + (\mathbf{u}\nabla)\nabla \cdot \mathbf{u} - (\nabla \cdot \mathbf{u})^2, \quad Q = \frac{\partial \mathbf{u}}{\partial T} \nabla D + (\mathbf{u}\nabla)(\mathbf{u}\nabla D). \tag{5.15}$$

In fact, we may choose the particle velocity at any depth as a physical variable, and recalculate ∇q_0 from (5.6); this leads to other forms of nonlinear dispersive equations for long waves (Wei et al. 1995, Madsen and Schaffer 1998, Agnon et al. 1999, Chen et al. 2000, Kim et al. 2003, Madsen et al. 2002, 2003). For most of them, the obtained linear dispersion relation has a Padé-polynomial form like (Madsen et al. 2003)

$$\frac{\Omega^2}{gDK^2} = \frac{1 + K^2 D^2/6 + K^4 D^4/120}{1 + K^2 D^2/2 + K^4 D^4/24}, \tag{5.16}$$

which is a very good approximation of the exact dispersion relation (2.52) in a relatively wide range of water depths KD (until depth of order $KD \approx 10$). Therefore, models of this type might be called fully nonlinear and dispersive systems of long waves (Boussinesq-like systems).

For the case of weakly nonlinear and weakly dispersive waves, all the Boussinesq-like models reduce to the Peregrine system (Peregrine 1967, 1972)

$$\frac{\partial \eta}{\partial T} + \nabla \cdot [(D+\eta)\mathbf{u}] = 0,$$

$$\frac{\partial \mathbf{u}}{\partial T} + (\mathbf{u}\nabla)\mathbf{u} + g\nabla \eta = \frac{D}{2}\frac{\partial}{\partial T}\left[\nabla(\nabla \cdot (\mathbf{u}D)) - \frac{D}{3}\nabla(\nabla \cdot \mathbf{u})\right]. \tag{5.17}$$

If the wave propagates mostly in one direction, and the bottom slope is small enough to neglect the wave reflection, the Peregrine system can be reduced to the famous Korteweg-de Vries and Kadomtsev-Petviashvili equations. At first, system (5.17) can be re-written in the form of a nonlinear wave equation for the water surface elevation, η,

$$\frac{\partial^2 \eta}{\partial T^2} - \nabla \cdot [C^2 \nabla \eta] = \Pi\{\eta, \mathbf{u}\} := -\nabla \cdot \left[\frac{\partial (\eta \mathbf{u})}{\partial T} + D\mathbf{F} - D(\mathbf{u}\nabla)\mathbf{u}\right], \tag{5.18}$$

where C is the long-wave speed, $C^2 = gD$. The function Π specifies nonlinear and dispersive terms that are weak. Let us introduce a new temporal variable

$$s = \tau(X,Y) - T, \tag{5.19}$$

where the function τ will be determined later. With these new variables, Eq. (5.18) reads

$$\left[1 - C^2(\nabla\tau)^2\right]\frac{\partial^2\eta}{\partial s^2} - \frac{\partial}{\partial s}\left[2C^2\nabla\tau\nabla\eta + \eta\nabla\cdot(C^2\nabla\tau)\right] - \nabla\cdot(C^2\nabla\eta) = \Pi. \tag{5.20}$$

When curvatures of the wave front and bottom slope are small (this assumption is normal for the ray theory), the elevation is a fast function of s and a slow function of spatial coordinates. Due to this, the last term on the LHS of Eq. (5.20) may be neglected. Owing to the weakness of nonlinearity and dispersion on the RHS of (5.20), a linear relation of long waves

$$\mathbf{u} = g\nabla\tau\eta \tag{5.21}$$

can be applied as follows from Eq. (5.17). As a result, Eq. (5.20) splits into a system of two equations for τ and η (see Engelbrecht et al. 1988 and Dingemans 1996)

$$(\nabla\tau)^2 = C^{-2}(X,Y) = (gD)^{-1}, \tag{5.22}$$

$$\frac{\partial}{\partial s}\left[2C^2\nabla\tau\nabla\eta + \eta C^2\Delta\tau + \eta\nabla\tau\nabla C^2\right] + \Pi\{\eta\} = 0, \tag{5.23}$$

The first Eq. (5.22) is the famous eikonal equation of the ray theory for long waves, allowing the determination of ray paths and wave fronts. This equation may be rewritten in the Hamiltonian form (3.6) (see Chap. 3 and discussion in Sect. 3.1). In the context of rogue waves, it determines the random location of caustics, where the wave field exhibits high amplitudes. The second Eq. (5.23), once integrated, results in

$$2C^2\nabla\tau\nabla\eta + \eta(C^2\Delta\tau + \nabla\tau\nabla C^2) + \frac{3\eta}{D}\frac{\partial\eta}{\partial s} + \frac{D}{3g}\frac{\partial^3\eta}{\partial s^3} = 0. \tag{5.24}$$

Noting that $\nabla\tau\nabla\eta = C^{-1}\partial\eta/\partial l$ and calculating $\Delta\tau = b^{-1}d(b/C)/dl$, where l is a coordinate along the ray and b is a distance between neighboring rays, then Eq. (5.24) gives the following equation (see Pelinovsky 1982, Dingemans 1996)

$$C\frac{\partial\eta}{\partial l} + \frac{3\eta}{2D}\frac{\partial\eta}{\partial s} + \frac{D}{6g}\frac{\partial^3\eta}{\partial s^3} + \frac{C\eta}{4Db^2}\frac{d(Db^2)}{dl} = 0. \tag{5.25}$$

This equation stands for the energy flux conservation (3.7) used previously for monochromatic waves in the linear approximation. Equation (5.25) governs the evolution of weakly nonlinear and weakly dispersive waves in a basin with variable depth. The first time, it was derived by Ostrovsky and Pelinovsky (1975). In basins of constant depth (5.25), it reduces to the Korteweg-de Vries equation

$$C\frac{\partial \eta}{\partial X} + \frac{3\eta}{2D}\frac{\partial \eta}{\partial s} + \frac{D}{6g}\frac{\partial^3 \eta}{\partial s^3} = 0, \quad s = \frac{X}{C} - T, \tag{5.26}$$

$$\frac{\partial \eta}{\partial T} + \frac{3C\eta}{2D}\frac{\partial \eta}{\partial X'} + \frac{CD^2}{6}\frac{\partial^3 \eta}{\partial (X')^3} = 0, \quad X' = X - CT, \tag{5.27}$$

and describes the evolution of an initial spatial disturbance. The Korteweg-de Vries equation is an etalon equation in the theory of nonlinear waves; it can be solved exactly with the help of the Inverse Scattering Technique (IST) (Novikov et al. 1984, Drazin and Johnson 1989).

When the wave field is directional with significant variation in the transversal direction, then the last term on the LHS of Eq. (5.20) can not be neglected and should be replaced by $\partial (C^2 \partial \eta / \partial Y)/\partial Y$, where Y is the transverse coordinate. This term does not allow integration of Eq. (5.23), and the modified evolution equation is now of fourth order, instead of the third order as is Eq. (5.25). In particular, for basins of constant depth, it becomes

$$\frac{\partial}{\partial X'}\left[\frac{\partial \eta}{\partial T} + \frac{3C\eta}{2D}\frac{\partial \eta}{\partial X'} + \frac{CD^2}{6}\frac{\partial^3 \eta}{\partial (X')^3}\right] + \frac{C}{2}\frac{\partial^2 \eta}{\partial Y^2} = 0, \quad X' = X - CT. \tag{5.28}$$

This equation is the famous Kadomtsev–Petviashvili equation that is also completely integrable (Novikov et al. 1984, Drazin and Johnson 1989).

These evolution equations for shallow water waves will be used in the next sections to study the freak-wave phenomenon.

5.2 Nonlinear-Dispersive Focusing of Unidirectional Shallow-Water Wave Fields

Unidirectional shallow water waves are known to be stable with respect to long perturbations. An initial wave field represented by weakly modulated wave trains evolves in time with some change of the shape of the trains, but the waves remain uniform, and their amplitudes do not vary significantly (Kit et al. 2000). Therefore, the modulational instability mechanism that is important for deep water cannot provide wave energy exchange and focusing within a wave group in shallow water. Dispersion, however, still may spawn rogue waves, although the shallow water dispersion law is different from that of deep water. The Korteweg-de Vries (KdV) equation (5.27), derived in the previous section, is a basic weakly dispersive and weakly nonlinear model. This equation was the first that exhibited exact soliton solutions (Zabusky and Kruskal 1965), and the associated Cauchy problem was integrated by using IST (Gardner et al. 1967). The soliton solution is a steady-state solution of Eq. (5.27)

$$\eta(X,T) = H\text{sech}^2\left[\sqrt{\frac{3H}{4D}}\frac{X - VT}{D}\right], \quad V = C\left[1 + \frac{H}{2D}\right], \tag{5.29}$$

corresponding to a moving solitary crest on the free surface that was first observed by Scott Russel in a narrow channel in 1844. The soliton length is formally infinite, but physically it is naturally determined at the level of elevation 0.5:

$$\lambda_s = 2D\sqrt{\frac{4D}{3H}}\ln(1+\sqrt{2}) \cong 2D\sqrt{\frac{D}{H}}. \tag{5.30}$$

For instance, a soliton of 1 m height has a length of about 60 km in water of 1 km depth. Solitons are generated from a wide class of initial disturbances that vanish at infinity. Its upper number may be estimated by the formula (Drazin and Johnson 1989)

$$N_s \leq 1 + \frac{3}{4D^3}\int_{-\infty}^{\infty}|X|(1+\mathrm{sgn}(\eta))\eta(X)dX. \tag{5.31}$$

The qualitative character of nonstationary processes of nonlinear wave dynamics within the framework of the Korteweg-de Vries equation can be clarified from the nondimensional form of Eq. (5.27)

$$\frac{\partial \zeta}{\partial t} + \zeta\frac{\partial \zeta}{\partial x} + \frac{1}{9Ur}\frac{\partial^3 \zeta}{\partial x^3} = 0, \tag{5.32}$$

where the dimensionless variables $\zeta = \eta/A_0$, $x = X/\lambda_0$, and $t = (3CA_0T)/(2\lambda_0 D)$ are normalized by the amplitude A_0 and length λ_0 of the initial disturbance, respectively. Here, Ur is the Ursell parameter

$$Ur = \frac{A_0\lambda_0^2}{D^3}. \tag{5.33}$$

The physical meaning of the Ursell parameter is evident: it characterizes the ratio of nonlinearity to dispersion. When the Ursell parameter is small, the nonlinearity can be neglected and the wave is a linear dispersive wave. Alternatively, if the Ursell parameter is large, dispersion can be neglected, and the wave evolves as a nonlinear nondispersive wave forming a steep front. For a soliton solution, $Ur = 4$, and this value is marginal, separating nonlinear nondispersive and linear dispersive regimes. This approach and exact solutions will be used in this section to investigate the effect of nonlinear-dispersive focusing.

To study rogue wave generation, it is convenient to invert the time variable in the evolution equation, as similarly done in Chaps. 3 and 4. To do this, the spatial coordinate, X, in the KdV equation should be replaced by $-X$. Hence, an initial value problem for an expected rogue wave may be considered to draw some inferences about wave fields that could form a freak wave (when time is reversed back to its normal run). Vanishing at infinity ($X \to \pm\infty$), boundary conditions result in the simplest analytical analysis of the Cauchy problem.

In particular, solutions for an initial wave in the form of a delta-function (singular initial data) can be obtained analytically (Drazin and Johnson 1989). According

5.2 Nonlinear-Dispersive Focusing of Unidirectional Shallow-Water Wave Fields

to the exact solution, a positive[1] delta-function evolves into a solitary wave (one soliton) and oscillating dispersive tail. The generated soliton is

$$\eta = \frac{3Q^2}{4D^3} \operatorname{sech}^2 \left[\frac{3Q}{4D^3} \left(X - C \left(1 + \frac{3Q^2}{8D^4} \right) T \right) \right], \qquad (5.34)$$

where Q is the delta-function intensity. The soliton moves with a larger speed and therefore is in front of the wave train; other waves are distributed in space according to the dispersion of the wave velocity. The soliton conserves its shape and energy, while the dispersive tail is spreading in space and thus vanishes. Therefore, the solitary part of the solution is the asymptotic solution of a Cauchy problem for the KdV equation. When the delta function is negative, only a dispersive tail may occur.

Bearing in mind that time may be reversed, this solution actually shows that a delta-function wave may be formed from weak-amplitude waves with or without a soliton. The KdV model does not limit the amplitude of possible abnormal waves; the wave-focusing mechanism due to dispersion is applicable in the nonlinear case as well, but the wave field structure is more complicated and includes amplitude-frequency modulated wave packets and solitons. This process was investigated in detail by Pelinovsky et al. (2000) and Kharif et al. (2000), and is shown in Fig. 5.1 (in the system of coordinates moving with speed C). The value of maximum wave amplitude in the domain increases rapidly and then decreases rapidly again (Fig. 5.2); this explains the short-lived character of rogue waves. Nevertheless, it should be emphasized that the Korteweg-de Vries model is a weakly nonlinear model, and use of singular initial conditions (like delta functions) may be nonphysical. Smoothed bell-like initial conditions with characteristic amplitude, A_0, and length, λ_0, may be considered as well. Negative initial disturbances result in a dispersive tail only; therefore, this process is qualitatively similar to the linear limit (see Sect. 3.2). In this case, the rogue wave is a deep hole on the sea surface (see Fig. 5.3). Positive initial pulse (a crest) may transform into solitons; their number and amplitudes depend on the Ursell parameter (5.33).

When the Ursell parameter is large, the amplitudes of generated solitons are comparable with the amplitude of the initial disturbance (in the limiting case $Ur \gg 1$, the amplitude of the leading soliton is two times larger than the initial pulse). Therefore, an initial pulse (that is supposed to be an expected rogue wave) cannot be considered a model of a freak wave, since condition (I.1) for the wave field amplitude amplification is not satisfied.

In the case of a small Ursell parameter, only one soliton is formed with a small amplitude (proportional to Ur). The initial pulse may now be much larger than the wave field at large time, since the soliton amplitude is small, and the dispersive train vanishes. When time is inversed, the evolution may represent a likely process of a rogue wave generation (see Fig. 5.1), while the pulse-like wave may be considered as a freak wave.

---

[1] This sign depends on the sign of the nonlinear coefficient in the Korteweg-de Vries equation, which is positive for surface water waves.

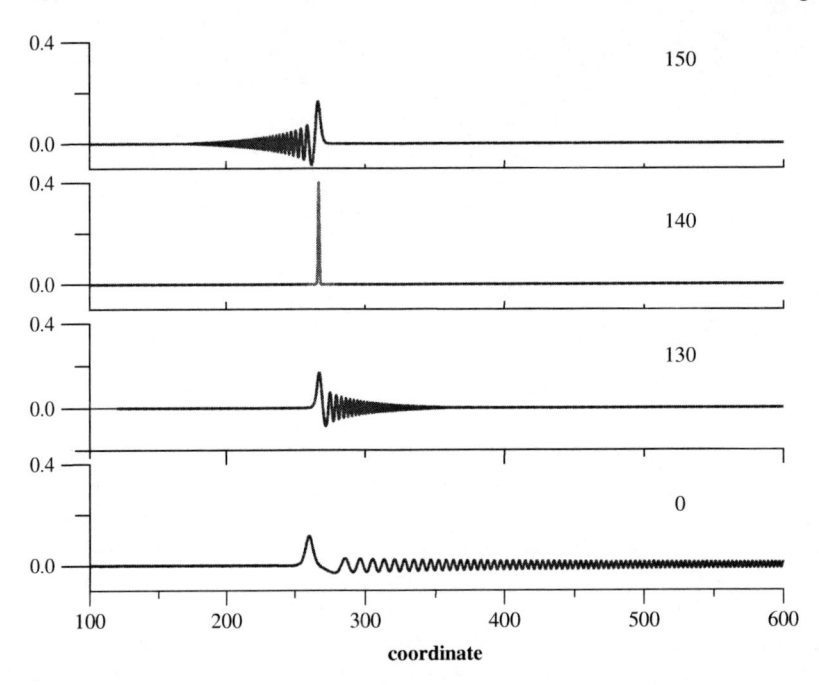

Fig. 5.1 Freak wave formation in shallow water. Numbers denote moments of time (*scaled*)

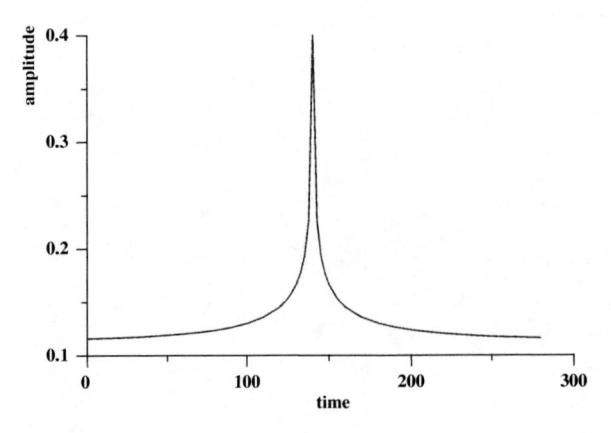

Fig. 5.2 Maximum wave amplitude versus time in the process of the freak wave formation given in Fig. 5.1

It is noteworthy to say that solitons do not play a crucial role in this freak wave generation scenario. The huge wave is mainly due to the frequency-modulated dispersive wave train.

The nonlinear-dispersive mechanism of freak wave formation is relatively robust; weak variation of the wave field parameters modifies the shape and amplitude of the freak wave, but is unable to prevent its occurrence. Specific numerical simulations have been performed in Pelinovsky et al. (2000) and Talipova et al. (2008) to

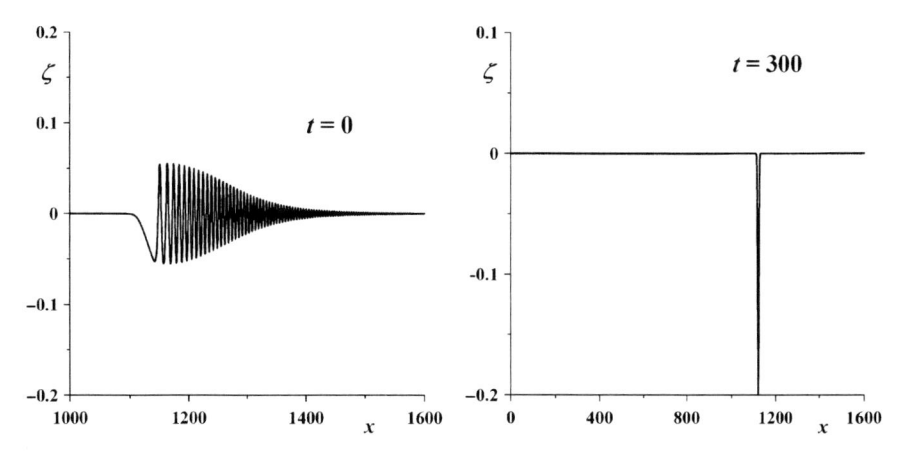

5.2 Nonlinear-Dispersive Focusing of Unidirectional Shallow-Water Wave Fields

Fig. 5.3 Generation of a deep hole in shallow water

highlight this property. A wave packet generated from a positive narrow pulse (as shown in Fig. 5.1) is inverted in space, $X \to -X$, and then several individual waves are canceled. This wave field is used to start the numerical simulation of the KdV equation. The simulation would result in the initially imposed positive impulse if the wave field has not been cut. If the soliton is deleted from the dispersed wave field (Fig. 5.4a), its focusing results in a freak wave with a large crest and following deep trough (a sign-variable wave; see Fig. 5.4b). If the first negative wave in the train shown in Fig. 5.4a is cut in addition to the soliton wave (Fig. 5.4c), the generated huge wave represents an almost positive pulse (a crest) with no deep neighboring troughs (Fig. 5.4d). The heights of computed abnormal waves in both cases satisfy the amplitude amplification criterion for rogue waves (I.1). Many natural observations support the existence of sign-variable rogue waves (see Chap. 1).

Similar analyses have been performed with a "solitonless" wave train, resulting from a negative pulse disturbance (see Fig. 5.3). If the leading negative oscillation is deleted (see Figs. 5.3a. and 5.5a), the huge wave is represented by several intense waves (Fig. 5.5b) that could be related to the observation of the "three sisters" also presented in Chap. 1.

Besides smooth solutions, singular exact solutions of the Korteweg-de Vries equation may be found (Matveev 2002). Similar to the soliton solutions, they preserve their identity, manifesting elastic collision with other waves. The positon solution is given as an example, although other solutions exist (negaton, singularities, a rational solution; see Matveev 2002)

$$\frac{\eta}{D} = -128p^2 \frac{\sin\Theta(\sin\Theta - p\Psi\cos\Theta)}{(\sin 2\Theta - 2p\Psi)^2}, \qquad (5.35)$$

where

$$\Theta = \frac{\sqrt{6}p}{D}\left(X - (1-4p^2)CT\right), \quad \Psi = \frac{\sqrt{6}}{D}\left(X - (1-12p^2)CT\right).$$

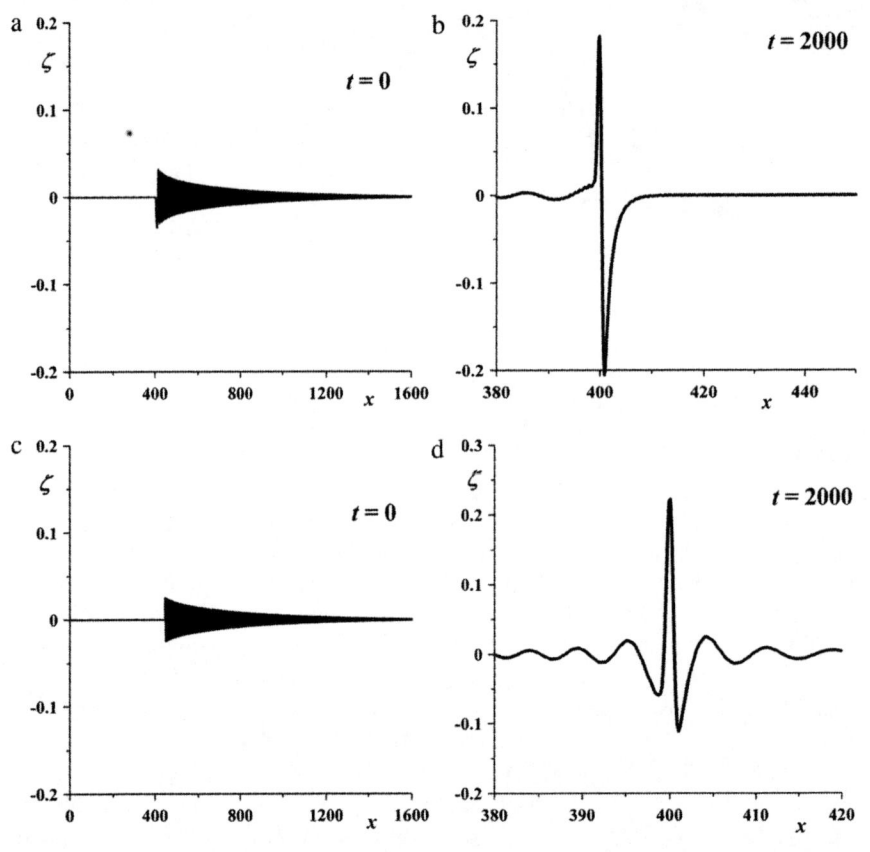

Fig. 5.4 "Non-optimal" focusing of the wave train in shallow water: initial conditions (**a, c**) and resulting waves (**b, d**)

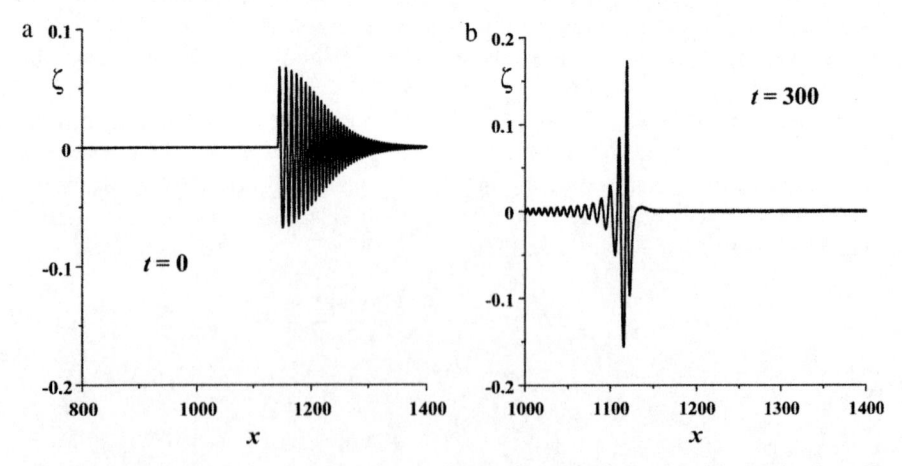

Fig. 5.5 Non-optimal generation of an abnormal wave from the wave train with negative "mass"

The parameter of this solution is p. A positon solution as a function of X has a second-order pole and, therefore, has an infinite energy; the tails of the oscillatory solution. Such solutions cannot be realized physically. They show, moreover, a tendency of smooth solutions of the KdV equation, close to waves with very high peaks.

The solution of the associated scattering problem with periodic boundary conditions is, in fact, much trickier to implement, since it operates with special theta functions. The detailed analysis of periodic solutions of the KdV equation is given in a series of papers by Osborne and coauthors (see, for instance, Osborne 1995, Osborne et al. 1998). The solution of the KdV equation is represented by a linear superposition of nonlinear oscillatory modes (multiple quasi-cnoidal waves) in the associated spectral problem. The freak wave in this approach is the superposition of these modes with suitable phases.

A statistical analysis of shallow-water rogue-wave characteristics has been conducted by Pelinovsky and Sergeeva (2006) with the help of direct numerical simulation of the KdV equation, with periodic boundary and random initial conditions; these results will be discussed in the next section. We would like to emphasize that a superposition of random and weak frequency modulated deterministic components still can efficiently spawn rogue waves, as it is shown in Fig. 5.6 (taken from Pelinovsky et al. 2000). So, freak waves in shallow water may be generated from a wide class of wave fields with the help of the nonlinear dispersive focusing.

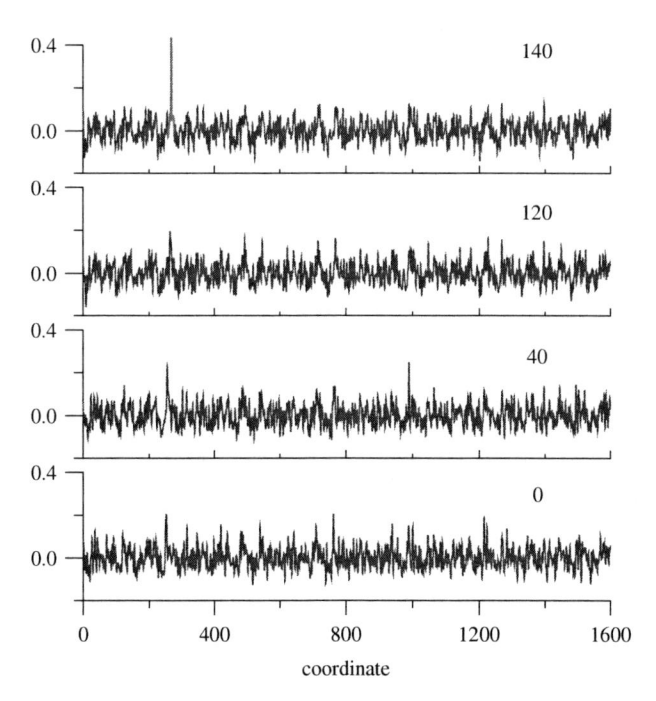

Fig. 5.6 Freak wave formation from the combination of a random field and frequency modulated wave train. Numbers denote moments of time (*scaled*)

5.3 Numerical Modeling of Irregular Wave Fields in Shallow Water (KdV Framework)

In the previous section, it has been shown, with the help of exact and numerical solutions of the KdV equation, how nonlinear-dispersive wave focusing may efficiently generate rogue waves. The KdV equation is integrable with the help of the IST, and this property supports the existence of solitons. When the wave field vanishes as coordinates tend to infinity, solitons are known to represent long-time asymptotic wave behavior, since quasilinear waves decay but solitons remain unchanged.

As shown above in the framework of the KdV equation, the nonlinear dispersive focusing of the wave trains is the major mechanism of freak wave occurrence. The random wave field is characterized by the modulation of the amplitude and frequency of waves. Therefore, the focusing mechanism should "work" in a random field. Meanwhile, the KdV equation is fully integrable, demonstrating an important role of the solitons in nonlinear wave dynamics. For initial disturbances vanishing at infinity, the solitons correspond to the final stage of the wave field evolution, and these results are well known. When the initial disturbance corresponds to the sine periodic wave, its evolution leads to soliton formation and its disappearance (recurrence phenomenon), as has been shown by Zabusky and Kruskal (1965).

Later, this process was investigated for different values of nonlinearity/dispersion ratio (the Ursell parameter given by (5.33)) and large times (see Salupere et al. 2002, 2003a,b and references therein). Actually, an initial sine state is not fully reconstructed at large time, and soliton ensembles play an important role in the long-time behavior of a nonlinear wave field, especially for large values of the Ursell parameter. The dynamics of the soliton ensembles, even for this simple initial sine condition, are very complicated and perhaps may be interpreted as solitonic turbulence, which can be considered as a combination of "rarefied solitonic gas" and the residue of oscillating quasilinear waves (Salupere et al. 1996).

Zakharov (1971) used the inverse scattering method to show that paired collisions occurring between solitons, and the interaction with a nonsoliton field, could not change the amplitude of the soliton. As a result, the total soliton velocity distribution function does not depend on time. In real situations of wind waves, the values of the Ursell parameters are not too large and the dispersive trains contribute significantly to the statistical wave characteristics.

Meanwhile, physically observed wave characteristics (spectra, amplitude, and height distributions) will change. The nonlinear energy exchange between different spectral components even for initial narrow-band wave fields is significant: a wave packet may split into several groups with different carrier wave numbers, and the wave profile becomes asymmetrical (Kit et al. 2000, Grimshaw et al. 2001, Groesen and Westhuis 2002). A wave realization, represented by multicnoidal waves and solitons, varies in space and time more significantly and its behavior is irregular (quasi-chaotic). Moreover, when an initial spectrum has two peaks, such a state is unstable (Zakharov 1971, Onorato et al. 2005), and therefore the wave dynamics should be complicated. As a result, the statistical moments and the distribution functions of the wave field change in time; its spatial spectrum also varies. Under

5.3 Numerical Modeling of Irregular Wave Fields in Shallow Water (KdV Framework)

the assumption of random initial conditions, the properties of such wave fields may be studied with the help of the random functions theory. In fact, we know only one mathematical paper (Murray 1978) where the soliton generation from irregular data is studied, but random wave characteristics have not been considered.

The direct numerical simulation of the KdV equation with periodic boundary conditions is applied in Pelinovsky and Sergeeva (2006) to study the statistical characteristics of wave fields and probability distributions of freak waves. In these simulations, the dimensionless form of the KdV equation (5.32) is used where normalization with significant wave amplitude (for random wave field the significant wave amplitude, A_s, is equal to 2σ, where σ^2 is the variance (2.72)), and carrier wave number K_0 (for random wave field it is the spectral peak wave number) are employed.

The numerical integration of the KdV equation (5.32) with periodic boundary conditions: $\zeta(0,t) = \zeta(L,t)$ is based on a pseudospectral method (Fornberg 1998). A zero-mean random wave field is described by a Fourier series containing 256 harmonics

$$\zeta(x,0) = \sum_{j=1}^{256} \sqrt{2S(k_j)\Delta k} \cos(k_j x + \varphi_j), \qquad (5.36)$$

where $S(k)$ is the initial nonsymmetric spectrum, $k_j = j\Delta k$, Δk is the sampling wavenumber, varying from 0.03 to 0.023, and the phase φ_j is a random variable, uniformly distributed in the interval $[0, 2\pi]$. The length of the initial realization is $L = 2\pi/\Delta k$. The initial spectrum is assumed to have a Gaussian shape of amplitude Q, and width δ:

$$S(k) = Q\exp\left(-\frac{(k-1)^2}{2\delta^2}\right). \qquad (5.37)$$

The parameter Q is chosen so that $\int_0^\infty 2S(k)\,dk = \sigma_0^2 = 1/4$ (σ_0^2 is the dimensionless variance). The spectral width parameter δ and the Ursell parameter both determine the dynamics of the nonlinear wave field. The sizes of the spectral domain (256 harmonics) and the characteristic spectrum widths are chosen to provide the spectrum decay when k is large. The initial spectra with a cut-off spectrum tail are presented in Fig. 5.7.

In numerical experiments by Pelinovsky and Sergeeva (2006), the Ursell parameter varies from 0.07 to 0.95, and the spectrum width varies from 0.27 to 0.18. Here, only the case $\delta = 0.27$ will be presented in detail. The statistical characteristics are computed for each time step and are averaged over 500 ensembles, which corresponds to a total wave record of about 15,000 individual waves to provide sufficient statistics. The computation is performed for relatively large time evolution, compared with the characteristic time scale of nonlinear effects (till $t = 100$) and includes about 1,000 wave periods depending on the initial conditions. This simulation time is sufficient for the manifestation of nonlinear and dispersive effects and to reach equilibrium conditions.

The evolution of a wave record is displayed in Fig. 5.8 for different instants of time. It is obviously seen that the wave profile becomes asymmetric, so that the

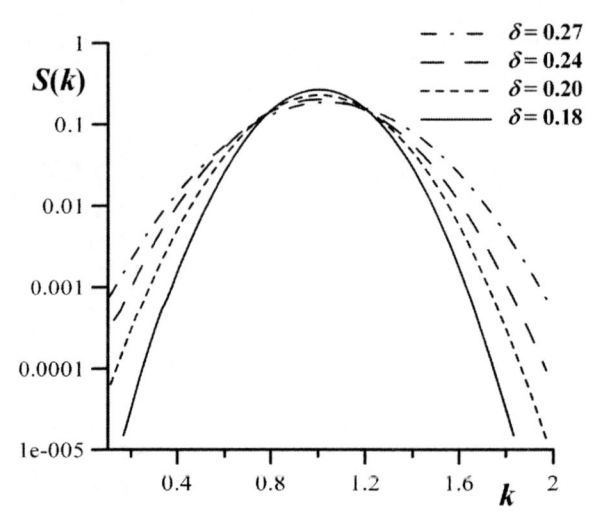

Fig. 5.7 Initial spectra for different widths δ

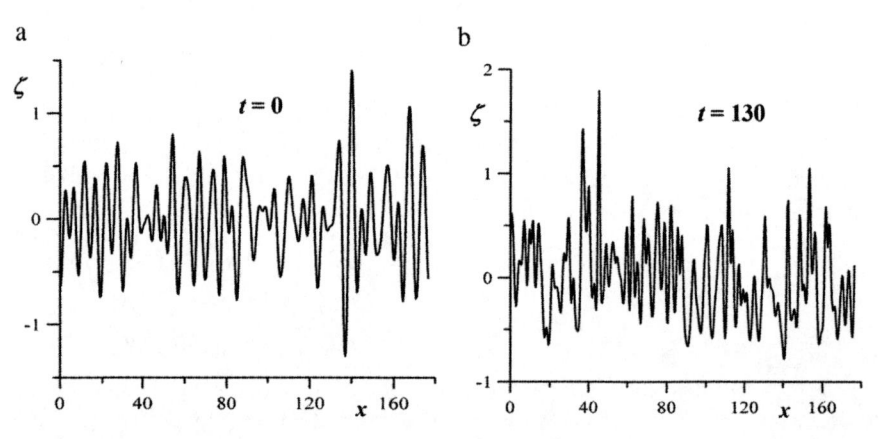

Fig. 5.8 Wave profiles at different instants of time ($Ur = 0.73$)

crests are sharp while the troughs are gentle. It is interesting to analyze the trajectory patterns (Fig. 5.9) presented in the time-space plane. This figure evidently shows solitons' traces for different initial conditions. The number of visible solitons, even for $Ur = 0.95$, is about 5; this means that solitons do not contribute significantly to the total random field. Under conditions of strong nonlinearity, the propagation gives rise to a maximum value of peak amplitude in most realizations (Fig. 5.10a). The key role of nonlinear effects in the formation of large wave amplitudes in this model becomes evident, as shown in Fig. 5.10b. This figure represents the distribution functions of the largest amplitudes, found for the case of numerical simulations and compared with the case of a linear propagation (when the nonlinear term in the KdV equation is canceled). Nonlinearity makes high amplitude wave occurrence more frequent.

5.3 Numerical Modeling of Irregular Wave Fields in Shallow Water (KdV Framework) 187

Fig. 5.9 Time-space plane of wave propagation for various Ur. Color gradations show the wave intensity. (**a**) The linear limit $Ur = 0$; (**b**) $Ur = 0.95$

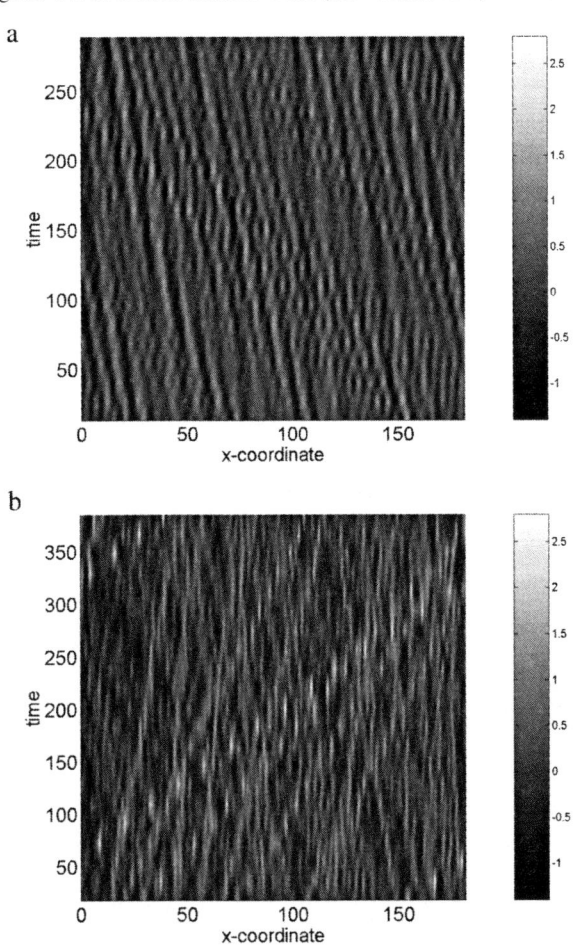

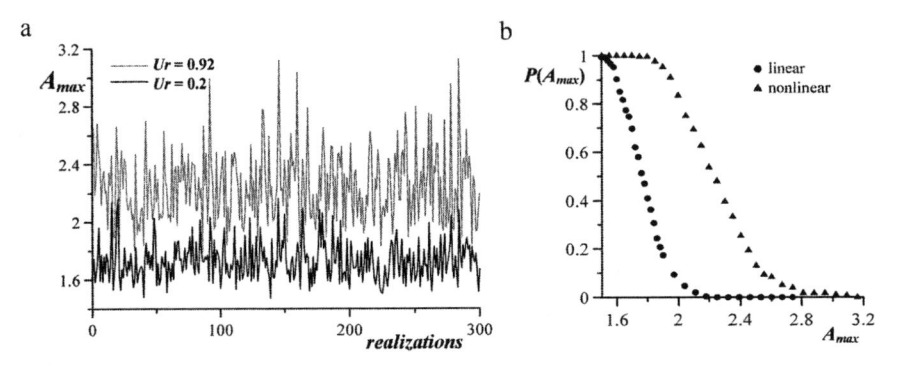

Fig. 5.10 Maximum of wave amplitudes in different realizations (**a**) and distribution of maximum crest amplitudes (**b**)

The first two statistical moments—the mean level and the variance—are integrals of the KdV equation, so that they remain unchanged during the process of wave evolution. The next two statistical moments define the skewness γ and kurtosis

$$\tilde{\kappa} = \kappa - 3, \qquad (5.38)$$

where γ and κ are defined by Eqs. (2.73) and (2.74) (see Chap. 2).

As known, the skewness is a statistical measure of the vertical asymmetry of the wave field. If the value of the skewness increases (positive), the crests are sharper, while the troughs are flatter. The kurtosis represents the degree of peakedness in the distribution and defines the contribution of large amplitude waves in the whole distribution. For a random Gaussian process, $\kappa = 3$, corresponding to $\tilde{\kappa} = 0$. When $\tilde{\kappa}$ is positive, the contribution of large waves is more significant. The computed evolution of statistical moments shows a stationary state existence and a transition to this state. The transition period is about 10-20 characteristic time scale of nonlinearity. During this process, both moments of the wave field tend to almost constant values (Fig. 5.11). Figure 5.12 displays the values of γ and $\tilde{\kappa}$, corresponding to this stationary mode.

For all conditions, the skewness is positive, and it means that the positive waves (crests) have larger amplitudes than the negative waves (troughs). The asymptotic value of skewness increases with an increase of the Ursell parameter; and therefore elevation (positive) waves are more visible in the nonlinear wave field than the depression (negative) waves. This conclusion corresponds to the known expressions for the classical cnoidal waves (sharp crest and flat trough).

The kurtosis tends to a negative asymptotic value for $Ur < 0.6$; therefore, the probability of large amplitude (freak) wave occurrence should be less than is predicted for Gaussian processes. For strong nonlinearity, the kurtosis asymptotic value exceeds zero, which indicates a high probability of large wave occurrence. Onorato

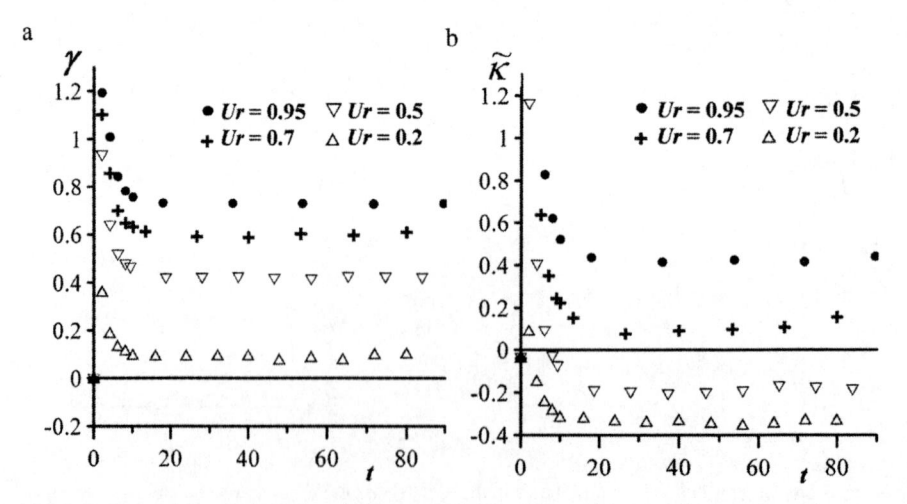

Fig. 5.11 Temporal evolution of statistical moments for different Ursell parameters

Fig. 5.12 Asymptotic value of the spectral moments as functions of the Ursell parameter Ur

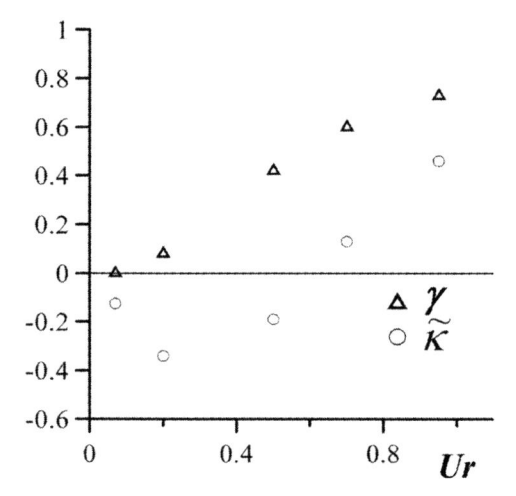

et al. (2001) and Tanaka (2001) showed by means of numerical experiments that for nonlinear random waves over deep water, the kurtosis $\tilde{\kappa}$ oscillates around some positive mean value. Janssen (2003) reports a positive fourth moment, calculated in the weak turbulence theory for deep-water waves that grow while the wave amplitude increases. Thus, the behavior of the fourth moment is qualitatively similar for strongly nonlinear waves in deep and shallow waters.

As expected due to nonlinearity, the spectrum evolves, widens, and tends to a stationary state (Fig. 5.13). This state, depending on the Ursell parameter, corresponds to the asymmetric wave shape; some energy is transferred to the low frequencies (spectrum downshift phenomenon). For large Ursell values, the spectral density is distributed almost uniformly at small k. The flatness of the spectrum is wider for $Ur = 0.95$ when the wave field is more energetic and nonlinear effects are more significant. The tendency to the flatness of the spectrum (Rayleigh-Jeans spectrum) is known for the statistical equilibrium with no sources and sinks.

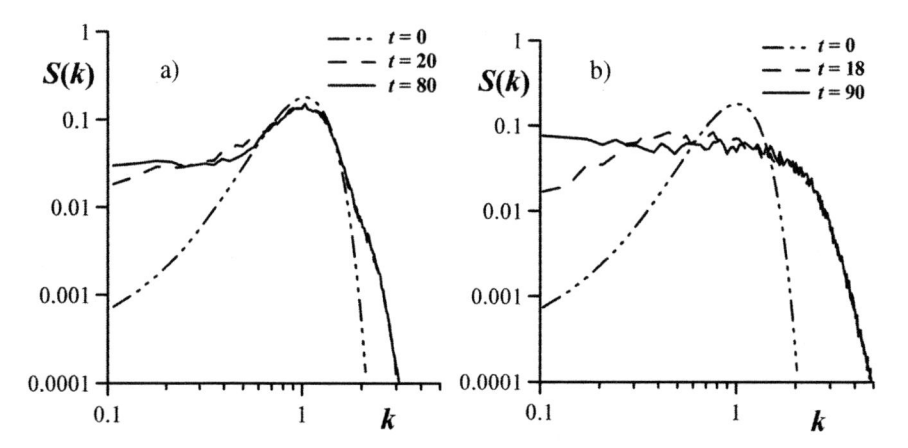

Fig. 5.13 Temporal evolution of spectra $S(k)$ for various Ur: (**a**) $Ur = 0.2$, (**b**) $Ur = 0.95$

It is important to mention that the spectrum downshifts into the low-frequency range even for an initial spectrum taken in a symmetric Gaussian form. For comparison, the downshift of the initial symmetric spectrum for deep-water waves is possible only in the extended version of the nonlinear Schrödinger equation, like the Dysthe equation, which includes an asymmetry of the wave field (Dysthe et al. 2003). The shallow water model based on the KdV equation is initially asymmetric due to the quadratic nonlinearity, and the asymmetry of the wave group is immediately obtained in the process of the wave evolution (Kit et al. 2000, Grimshaw et al. 2001, Groesen and Westhuis 2002). As already noticed, the spectrum becomes asymmetric with weak shifting in the short-wave range. For larger dimensionless wavenumbers $k(0.1 < k < 0.2)$, the spectrum may be approximated by the power law asymptotics $k^{-\alpha}$, where the slope of the spectrum, α, decreases with an increase of the Ur parameter (from $\alpha = 3.7$ for $Ur = 0.5$ till $\alpha = 2$ for $Ur = 0.95$).

The distribution of the wave crest amplitudes, calculated as a maximum between two zero-crossings, is presented in Fig. 5.14. For $Ur < 0.3$, the probability of small amplitudes ($A < 1.2$) exceeds the Rayleigh distribution, which is the theoretical approximation of a linear narrow-band Gaussian process (see Chap. 2); meanwhile, in the range of high amplitudes ($A > 1.5$), the distribution lays below the theoretical curve. For the more energetic wave field ($Ur > 0.3$), the asymptotic distribution exceeds the Rayleigh distribution, and the probability of the highest crest occurrence increases. In a qualitative sense, the shape of the amplitude distribution function does not contradict the behavior of the skewness and kurtosis (Fig. 5.12). The first one shows that positive waves have larger amplitudes than negative waves, whereas the second one indicates a significant contribution of the small waves in the whole distribution. Finally, these results allow us to estimate the probability of the rogue-wave occurrence (its amplitude exceeds twice the significant amplitude; see (I.1)). Freak waves should appear more frequently when the wave field is strongly nonlinear (high values of the Ursell parameter).

The same results are obtained when using experimental spectra of shallow-water waves in the coastal zone of the North Sea and in Lake Georgia in Australia

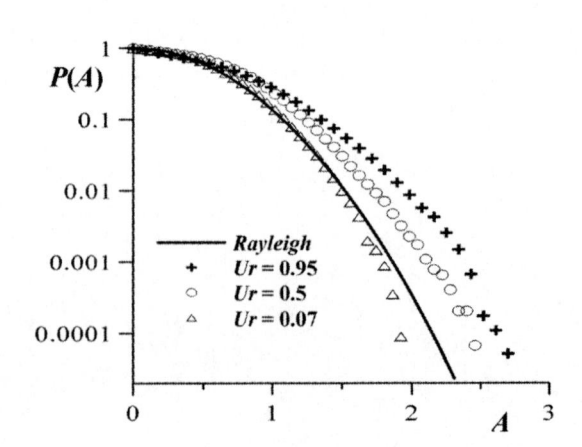

Fig. 5.14 Asymptotic crest amplitude distribution for different Ur numbers. *Solid line* corresponds to the Rayleigh distribution

(Kokorina and Pelinovsky 2005). The computed results confirm that the irregular nonlinear wave field does not satisfy the Gaussian statistics, and its statistical characteristics depend on the Ursell parameter, which represents the "ratio" of nonlinear to dispersive effects.

In this way, it is demonstrated that the nonlinear shallow water-wave field becomes asymmetric with sharp crests and flat troughs, which leads to a positive third statistical moment. The skewness grows monotonously with the increase of the Ur number. The behavior of the 4th statistical moment (kurtosis $\widetilde{\kappa}$) is nonmonotonic. It is negative when $Ur < 0.8$, which indicates a significant contribution of small amplitude waves to the total distribution. When the initial disturbance is more nonlinear, then the kurtosis exceeds the zero level, at which it increases with a growth of Ur. For small Ur numbers, close to zero, the probability distribution function slightly deviates from the theoretical Rayleigh distribution. For $Ur > 0.3$, the computed curve lies above the theoretical distribution, which means a higher probability of large wave formation—namely freak-wave occurrence. An important result is the existence of a steady state for statistical characteristics: statistical moments (skewness and kurtosis), distribution functions, and spectral density. The computations demonstrate that both the statistical moments and distribution functions evolve until some bound level is reached. The analysis of a random wave-spectrum evolution shows the same effect. The initially symmetric power spectrum with a Gaussian shape broadens with time, and energy is transferred down the spectrum. For a period of time approximately equal to 20 of a characteristic time scale of nonlinearity, the spectrum relaxes to some stationary state with energy concentration in the low frequency range, as has been already noticed. The parameters of the equation—in particular the Ur parameter—influence the width of the steady spectrum. For strong nonlinearity, the established stable spectrum is wider, and the energy is distributed almost uniformly in the range of long waves.

5.4 Three-Dimensional Rogue Waves in Shallow Water

When two horizontal coordinates are considered, rogue waves can appear owing to (i) the focusing of transient wave groups, and (ii) spatial (geometric) focusing of water waves. Nonlinear models of spatially inhomogeneous wave fields are complex even in basins of constant depth. They have been used to model freak-wave occurrence in 3D transient trains. Qualitatively, nonlinear processes support linear mechanisms of huge wave formation (see references in Sect. 3.2).

To clarify new, essentially nonlinear effects occurring in spatial inhomogeneous wave fields, let us first consider the interaction of two oblique propagating solitary waves. Basic equations for weakly nonlinear and weakly dispersive water waves were discussed in Sect. 5.1. It is convenient here to rederive such equations for the "equivalent" potential, q (it corresponds to the dimensional first term in the series (5.6), q_1) and dimensionless water displacement, ζ (Miles 1977a,b, Pelinovsky 1996)

$$\zeta = -\frac{\partial q}{\partial t} - \frac{\alpha}{2}(\nabla q)^2 + \frac{\beta}{2}\frac{\partial^3 q}{\partial t^3}, \tag{5.39}$$

$$\frac{\partial^2 q}{\partial t^2} - \Delta q = -\alpha \frac{\partial}{\partial t}\left[\frac{1}{2}\left(\frac{\partial q}{\partial t}\right)^2 + (\nabla q)^2\right] + \frac{\beta}{3}\frac{\partial^4 q}{\partial t^4}, \tag{5.40}$$

where coordinates are normalized with the wave length, λ, time, with the wave period, displacement, and with the wave amplitude, A. As a result, two parameters—$\alpha = A/D$ and $\beta = (D/\lambda)^2$—characterize the weak nonlinearity and dispersion, respectively. When two solitons propagate in different directions, it is convenient to make a change of coordinates as follows:

$$\xi_1 = y\cos\Psi_1 + x\sin\Psi_1 - t, \quad \xi_2 = y\cos\Psi_2 + x\sin\Psi_2 - t, \quad \tau = \alpha t. \tag{5.41}$$

In these new variables, Eqs. (5.39) and (5.40) become

$$\zeta = (\partial_1 + \partial_2 - \alpha\partial_\tau)q - \alpha\left[\frac{(\partial_1 q)^2 + (\partial_2 q)^2}{2} + (1-2\theta)\partial_1 q \partial_2 q\right] - \frac{\beta}{2}(\partial_1 + \partial_2)^2 q, \tag{5.42}$$

$$\alpha(\partial_1 + \partial_2)\left\{2\partial_\tau q + \left[\frac{3}{2}(\partial_1 q)^2 + \frac{3}{2}(\partial_2 q)^2 + (3-4\theta)\partial_1 q \partial_2 q\right]\right\}$$
$$+ \frac{\beta(\partial_1 + \partial_2)^3 q}{3} - 4\theta \partial_1 \partial_2 q = 0, \tag{5.43}$$

where $\theta = \sin^2[(\Psi_1 - \Psi_2)/2]$ corresponds to the difference in the soliton propagation directions; ∂_1 and ∂_2 denote derivation with respect to coordinate ξ_1 and ξ_2, respectively. In particular, the case $\Psi_1 = 90°$ and $\Psi_2 = -90°$ corresponds to the counter propagation of solitary waves. The solution of Eq. (5.43), to the first order of the nonlinear parameter (assuming $\alpha \sim \beta$), can be sought as

$$q = F_1(\xi_1, \tau) + F_2(\xi_2, \tau) + \alpha F_{12}(\xi_1, \xi_2, \tau). \tag{5.44}$$

Here, $\partial F_{1,2}/\partial \xi_{1,2}$ (it is proportional to the water displacement in the linear theory of long waves) are the "non-interacting" solitons described by the unidirectional KdV equation

$$2\alpha \frac{\partial F_{1,2}}{\partial \tau} + \frac{3\alpha}{2}\left(\frac{\partial F_{1,2}}{\partial \xi_{1,2}}\right)^2 + \frac{\beta}{3}\frac{\partial^3 F_{1,2}}{\partial \xi_{1,2}^3} = 0. \tag{5.45}$$

After substitution of Eq. (5.44) in Eq. (5.43), and taking into account Eq. (5.45), the first nonlinear correction to the potential is expressed by

$$F_{12}(\xi_1, \xi_2, \tau) = \frac{3-4\theta}{4\theta}\left(\frac{\partial}{\partial \xi_1} + \frac{\partial}{\partial \xi_2}\right) F_1(\xi_1, \tau) F_2(\xi_2, \tau). \tag{5.46}$$

As a result, the series (5.44) can be written with the same accuracy as Miles (1977a,b)

$$q = F_1(\xi_1 + \rho_2, \tau) + F_2(\xi_2 + \rho_1, \tau), \tag{5.47}$$

5.4 Three-Dimensional Rogue Waves in Shallow Water

where

$$\rho_{1,2} = \alpha \left(\left(\frac{3}{4\theta} - 1 \right) F_{1,2}(\xi_{1,2}, \tau) \right). \tag{5.48}$$

Similarly, the water displacement at the first order of nonlinearity is given by

$$\zeta = N_1(\xi_1 + \rho_2, \tau) + N_2(\xi_2 + \rho_1, \tau) + \alpha I N_1 N_2, \tag{5.49}$$

$$N_i = \left(\frac{\partial}{\partial \xi_i} - \frac{\beta}{3} \frac{\partial^3}{\partial \xi_i^3} \right) F_i + \frac{\alpha}{4} \left(\frac{\partial F_i}{\partial \xi_i} \right)^2, \quad I = \frac{3}{2\theta} - 3 + 2\theta. \tag{5.50}$$

The result of the interaction of two solitons depends on the angle between the soliton directions (expressed by the parameter θ). The coefficient of the interaction is $I = 0.5$ for solitons propagating toward each other ($\Psi_1 - \Psi_2 = 180°$), then it weakly decreases (down to 0.464) when $\Psi_1 - \Psi_2$ decreases to 138°, and then grows to infinity when the waves copropagate.

The breakdown of the perturbation technique for waves propagating in almost the same directions is evident from the mathematical point of view, because the two new coordinates, ξ_1 and ξ_2, are not independent in this case. From a physical point of view, almost parallel propagation of two solitons leads to strong interaction between them, and each soliton changes the trajectory of the propagation of the other soliton. In the vicinity of the almost parallel wave propagation, the solution should be obtained directly from the nonlinear evolution equations: the Kadomtsev-Petviashvili equation if the waves propagate almost parallel, or the KdV equation if the waves propagate in one direction.

The Kadomtsev-Petviashvili equation was derived in Sect. 5.1 and is reproduced here in dimensionless form

$$\frac{\partial}{\partial x} \left(\frac{\partial \zeta}{\partial t} + 6\zeta \frac{\partial \zeta}{\partial x} + \frac{\partial^3 \zeta}{\partial x^3} \right) = -3 \frac{\partial^2 \zeta}{\partial y^2}, \tag{5.51}$$

where $\zeta = 3\eta/2D$, $x = X/D$, $y = Y/D$ and $t = CT/6D$. The Kadomtsev-Petviashvili equation is also integrable (Drazin and Johnson 1989) and therefore exact solutions can be used to study the soliton interaction. It is convenient to use the Hirota transformation

$$\zeta = 2 \frac{\partial^2}{\partial x^2} \ln \Gamma(x, y, t), \tag{5.52}$$

to reduce Eq. (5.51) to bilinear form

$$\Gamma \left(\frac{\partial^2 \Gamma}{\partial t \partial x} + \frac{\partial^4 \Gamma}{\partial x^4} + 3 \frac{\partial^2 \Gamma}{\partial y^2} \right) - \frac{\partial \Gamma}{\partial t} \frac{\partial \Gamma}{\partial x} - 3 \left(\frac{\partial \Gamma}{\partial y} \right)^2 - 4 \frac{\partial \Gamma}{\partial x} \frac{\partial^3 \Gamma}{\partial x^3} + 3 \left(\frac{\partial^2 \Gamma}{\partial x^2} \right)^2 = 0. \tag{5.53}$$

The plane soliton of the Kadomtsev-Petviashvili equation

$$\zeta = \frac{k^2}{2} \operatorname{sech}^2(k\xi/2), \quad \xi = k(x - py - Vt), \quad V = k^2 + 3p^2 \tag{5.54}$$

in the framework of Eq. (5.53) is expressed in the simple form

$$\Gamma = 1 + \exp(\xi). \tag{5.55}$$

Here, p determines the slope of the soliton trajectory in space. Similarly, the two-soliton solution of Eq. (5.53) can be written explicitly (Satsuma 1976):

$$\Gamma = 1 + \exp(\xi_1) + \exp(\xi_2) + r^2 \exp(\xi_1 + \xi_2), \tag{5.56}$$

$$\xi_i = k_i(x - p_i y - V_i t), \qquad r^2 = \frac{(k_1 + k_2)^2 + (p_1 + p_2)^2}{(k_1 - k_2)^2 - (p_1 + p_2)^2}.$$

Solitons are separated in space except the area of interaction around the moving point:

$$x^* = \frac{V_1 p_2 - V_2 p_1}{p_2 - p_1} t, \qquad y^* = \frac{V_1 - V_2}{p_2 - p_1} t. \tag{5.57}$$

The shapes of the large-amplitude waves occurring in the process of the two-soliton interaction for various angles between soliton fronts are given in Fig. 5.15 from the paper by Peterson et al. (2003). The wave amplitude depends strongly on the angle between the soliton fronts. Similar combinations of nonlinearly interacting waves may be often observed in nature near the coast (see Fig. 5.16).

To show the main features of the oblique interaction of solitons and calculate possible parameters of the enhanced wave, let us consider two solitons with the same amplitudes ($k_1 = k_2$) traveling symmetrically with respect to the Ox axis ($p_1 = -p_2$). As often used in wave physics, such an interaction is equivalent to the wave reflection at the wall located at $y = 0$. Then the condition $p_1 = -p_2$ has the meaning of

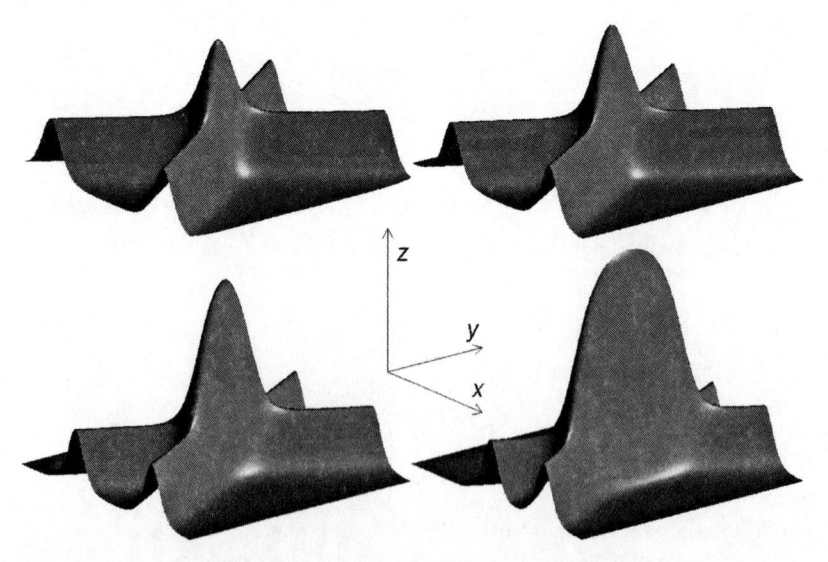

Fig. 5.15 Large-amplitude waves occurring in the process of soliton interaction. Reproduced from (Peterson et al. 2003) by permission of European Geosciences Union

5.4 Three-Dimensional Rogue Waves in Shallow Water

Fig. 5.16 Crescent nonlinear wave trains near the shore. A growing breaking wave is readily observed (Courtesy of I.I. Didenkulova)

the well-known Snell law (the reflection angle is equal to the incident angle). Such a situation with oblique soliton reflection is very often reproduced in laboratories (Melville 1980, Funakoshi 1980, Mase et al. 2002). In this case, the solitons propagate with the same speed ($V_1 = V_2$) and the pattern of wave interaction is stationary, while the interacting area moves along the Ox axis with constant speed. Under these conditions, the wave field is expressed as

$$\zeta(x,y,t) = 2k^2 \frac{1 + r\cosh[k(x-Vt)]\cosh(kpy)}{\{\cosh[k(x-Vt)] + r\cosh(kpy)\}^2}, \quad r = \sqrt{1 - \left(\frac{k}{p}\right)^2}. \quad (5.58)$$

The water displacement on the wall ($y = 0$) can be found from Eq. (5.58); and in dimensional variables it reads

$$\frac{H_w}{H_0} = \frac{4}{1 + \sqrt{1 - \frac{3H_0}{D\tan^2\Theta}}}, \quad (5.59)$$

where H_0 is the height of the incident soliton, and Θ is the angle between the soliton front and the Oy axis (see sketch in Fig. 5.17a).

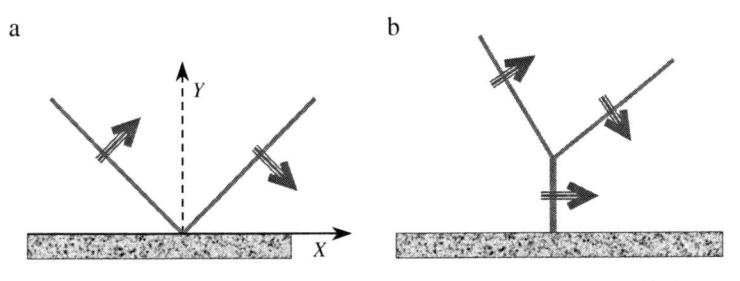

Fig. 5.17 Soliton reflection from a wall: quasi-linear reflection (**a**) and Mach stem formation (**b**)

At $\Theta \sim \pi/2$ (normal approach of the wave to the wall or, alternatively, counter soliton propagation), the wave height is increased almost twice and the same result can be obtained from the perturbation analysis (5.49). At small angles ($\tan^2 \Theta \sim 3H_0/D$), when the soliton propagates almost along the wall, the wave amplification near the wall can reach the value of four, and this is the result of joint action of nonlinear and diffraction effects that are of the same order of magnitude. But when the angle is very small ($k > p$), the solution (5.58) becomes complex and cannot describe the physical wave field. This means that wave fields at small angles are not stationary, and the interaction area should "take off" from the wall. In fact, this can be achieved from (5.58). When the solitons propagate toward each other with almost parallel wave crests, the incident and reflected solitons are well-separated everywhere in space (Fig. 5.17a). When $p \to k$, the induced soliton appears near the wall and propagates along the wall (Fig. 5.17b). The amplitude of this wave (5.59) and its speed (5.57) are different from those of a Korteweg-de Vries soliton, and it can be called a "virtual" soliton (Onkuma and Wadati 1983). Only under special conditions can this wave become a true soliton and propagate along the wall (the so-called Mach stem). Let us assume that the parameters of the incident (i) and reflected (r) solitons are related as

$$k_i + k_r = p_i + p_r. \tag{5.60}$$

Thus, $r = 0$ and the two-soliton solution (5.56) is

$$\Gamma = 1 + \exp(\xi_i) + \exp(\xi_r). \tag{5.61}$$

The wave (the Mach stem) propagates along the wall ($y = 0$) if

$$k_i p_i = k_r p_r, \tag{5.62}$$

which is the Snell law for wave reflection. Parameters of the reflected soliton may be found explicitly from Eq. (5.60) to Eq. (5.62)

$$k_r = p_i, \quad p_r = k_i > p_i, \tag{5.63}$$

and soliton speeds are not equal: $V_r < V_i$. It confirms that the process of wave reflection is not stationary and can be interpreted as a resonant interaction of three solitons: incident, reflected, and the Mach stem. The wave height at the wall can be found in Eq. (5.61), and in dimensional form it reads

$$\frac{H_w}{H_0} = \left[1 + \frac{\tan \Theta}{(3H_0/D)^2}\right]^2. \tag{5.64}$$

Formulas (5.59) and (5.64) describe the nonmonotonic character of the wave amplification. Its maximum (four) is achieved when the angle between waves is of the same order as the nonlinear parameter A/D. Formation of the Mach stem

5.4 Three-Dimensional Rogue Waves in Shallow Water

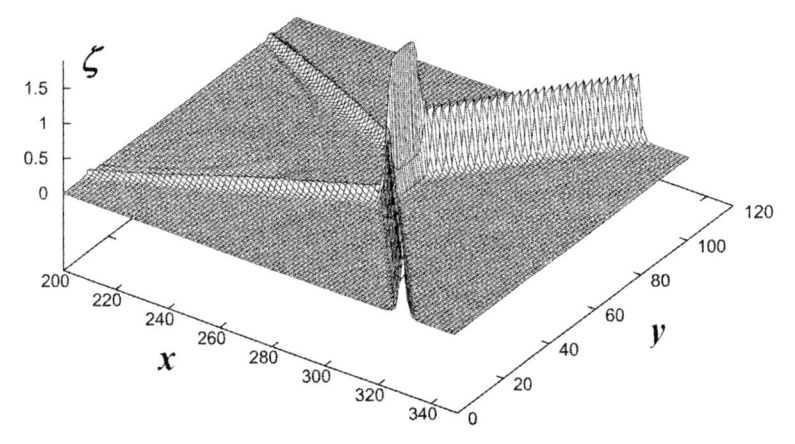

Fig. 5.18 Formation of the Mach stem in almost collinear soliton interaction. Reproduced from Porubov et al. (2005) with permission from Elsevier

was studied numerically by Porubov et al. (2005); Fig. 5.18 illustrates this process in the general case. The wave steepness in the process of two-soliton interactions can be enhanced to a value eight times that of the initial steepness (Soomere and Engelbrecht 2005).

It is important to note that two-soliton interaction leads to the formation of a rogue wave with an infinite lifetime. Specific numerical simulations of the Cauchy problem for the Kadomtsev-Petviashvili equation performed in Porubov et al. (2005) show that the result of the interaction of initially separated solitons depends strongly on the curvature of the initial fronts, and the maximum amplification in the interacting area can be very large. In fact, a combination of two different effects takes place in this case: nonlinear interaction of solitons and geometrical focusing. The same effect may be observed at random wave incidence (Mase et al. 2002). Figure 5.19 shows the effect of the Mach stem formation in a laboratory tank.

So, comparison with unidirectional wave-field dynamics in shallow water shows that soliton interactions play a significant role in localized rogue wave formation. Toffoli et al. (2006) performed detailed calculations of the statistical properties of shallow water waves in crossing seas within the framework of the Kadomtsev-Petviashvili equation. Numerical simulations indicate that the interaction of two noncollinear wave trains generates steep and high amplitude peaks, thus enhancing the deviation of the surface elevation from the Gaussian statistics. These peaks yield a modification of the upper tail of the probability density function for surface elevation, which significantly deviates from the distribution of wave elevation in the unimodal condition. The coexistence of two spectral peaks, therefore, enhances the nonlinearity of the wave field, which results in an increase of the skewness and kurtosis. Whereas this enhancement is negligible for nearly collinear waves, the skewness and kurtosis reach high values when the two spectral peaks have well-separated directions.

Fig. 5.19 Formation of Mach stem (see arrows) near the vertical wall at random wave incidence. Reproduced from (Mase et al. 2002) with permission from Elsevier

5.5 Anomalous High Waves on a Beach

The rogue wave phenomenon is usually discussed in terms of waves in seas and oceans far from the shores. Such unusual waves are observed also in the coastal zone and on coastlines. Excellent photos of freak waves on rocky coasts are given in Chap. 1 (Fig. 1.1h), when a freak wave reached height of 25 m approximately 4 sec after it became visible near the coast of Vancouver Island, Canada. Other freak waves attacked the breakwater in Kalk Bay (South Africa) on April 21, 1996 and August 26, 2005. In both events, the freak wave washed off the breakwater people, some of whom were injured. The freak waves induced panic at Maracas Beach (Trinidad Island, Lesser Antilles) on October 16, 2005, when a series of towering waves, many more than 25 feet high (height of 8 m), flooded the beach, forcing sea-bathers, venders, and lifeguards to run for their lives (see Fig. 1.1g).

The wave field in coastal zones contains strong coherent components and may be represented as the nonlinear superposition of solitary (solitons), cnoidal, and breaking waves (undular and smooth bores). Their interaction can generate narrow "spots" of large-amplitude freak waves. The bottom topography plays a significant role in spatial (geometric) interference of waves, resulting in the formation of random focusing and caustic points, where the wave field is amplified. The effect of water wave amplification in the coastal zone is well known. It means that probability of large-amplitude waves should increase in the coastal zone. In this section, we

5.5.1 Waves at Vertical Walls

First, one of the typical nonlinear effects in the coastal zone will be considered and analyzed when the wave propagates close to vertical walls (rocks, breakwaters, other vertical structures) and may suffer reflection. A simplified geometry of the coastal zone is shown in Fig. 5.20. The wave approaches the vertical wall located at $X = 0$ from the left. For the sake of simplicity, the incident wave is represented as a single crest, but later we will consider the incident wave as a continuous function, describing random crests and troughs. The basic equations for water waves in shallow water are

$$\frac{\partial \eta}{\partial T} + \frac{\partial}{\partial X}[(D+\eta)u] = 0, \qquad \frac{\partial u}{\partial T} + u\frac{\partial u}{\partial X} + g\frac{\partial \eta}{\partial X} = 0, \qquad (5.65)$$

where $u(X,T)$ is the depth-averaged horizontal velocity of the water flow (see Eq. (5.10)) and $\eta(X,T)$ is the vertical displacement of the sea level.

The boundary condition on the vertical wall corresponds to the total reflection of the wave energy and no penetration of fluid through the wall is considered:

$$u(X = 0, T) = 0. \qquad (5.66)$$

Another condition that concerns the approach of the incident wave to the wall from the left will be discussed. To solve Eq. (5.65), it is convenient to introduce the Riemann invariants

$$I_\pm = u \pm 2\left[\sqrt{g(D+\eta)} - \sqrt{gD}\right], \qquad (5.67)$$

and rewrite system (5.65) in the following form

$$\frac{\partial I_\pm}{\partial T} + C_\pm \frac{\partial I_\pm}{\partial X} = 0, \qquad (5.68)$$

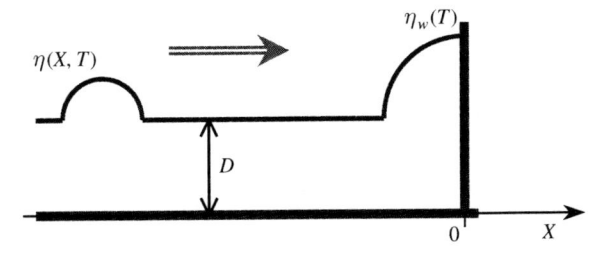

Fig. 5.20 Definition sketch of the considered geometry

where the characteristic speeds are

$$C_\pm = \pm\sqrt{gD} + \frac{3}{4}I_\pm + \frac{1}{4}I_\mp. \qquad (5.69)$$

According to Eq. (5.68) each invariant remains constant along the characteristic curves

$$\frac{dI_\pm}{dT} = 0 \quad \text{along} \quad \frac{dX}{dT} = C_\pm. \qquad (5.70)$$

Note that the characteristic speeds depend on both invariants, and nonlinearity bends the characteristics in the vicinity of the wall area where the incident and reflected waves interact. When taking into account conservation of the Riemann invariants, the effect of wave interaction yields phase corrections of the travel times of different parts of the wave profile. As a result, the water displacement at the vertical wall $\eta_w(T) = \eta(X=0,T)$ depends on the incident wave in a very complicated manner, and cannot be found in an explicit form. Nevertheless, the relation between values of the wave in the incident field and in the near-wall water oscillations can be derived explicitly. Outside the interaction near-wall area, the incident and reflected waves propagate independently. The incident wave is characterized by

$$I_- = 0, \quad u = 2\left[\sqrt{g(D+\eta)} - \sqrt{gD}\right], \quad I_+ = 4\left[\sqrt{g(D+\eta)} - \sqrt{gD}\right]. \qquad (5.71)$$

Due to the boundary condition (5.66), the incident invariant at the wall is

$$I_+ = 2\left[\sqrt{g(D+\eta_w)} - \sqrt{gD}\right]. \qquad (5.72)$$

From the conservation of I_+ along the characteristic curves it follows that

$$\frac{\eta_w(T)}{D} = 4\left[1 + \frac{\eta(T-\tau)}{D} - \sqrt{1 + \frac{\eta(T-\tau)}{D}}\right]. \qquad (5.73)$$

So, the water level on the wall can be expressed through the water displacement of the incident wave. Unfortunately, this method cannot predict the time-lag, τ, which is generally an unknown functional of the wave field in the interaction zone. This is why expression (5.73) cannot be straightforwardly applied for calculations of the water level oscillations near the vertical wall, even when all the characteristics of the incident wave are known. However, a practical formula can be derived from (5.73)—it is the relation between the extreme values of the wave field

$$\frac{R}{D} = 4\left[1 + \frac{A}{D} - \sqrt{1 + \frac{A}{D}}\right], \qquad (5.74)$$

where A is the positive or negative amplitude (crest or trough height) of the incident wave, and R is the amplitude of water level oscillations on the wall. This relation is plotted in Fig. 5.21 (solid line).

5.5 Anomalous High Waves on a Beach

Fig. 5.21 Amplitude of water oscillations at the wall (R) versus the incident wave amplitude (A) according to the linear (*dashed*) and nonlinear (*solid*) theories

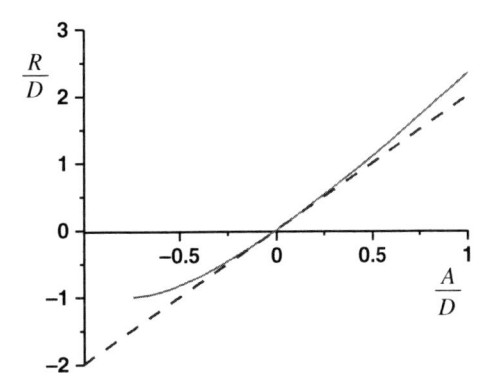

The linear limit gives the following relation between the wave characteristics, $R = 2A$. This curve is plotted by dashed line in Fig. 5.21 for comparison. As can be seen, the nonlinearity increases the crest height and decreases the trough height at the wall. In fact, the weak increase of the wave height, when the positive wave (crest) comes near the wall, was analyzed earlier by Mirchina and Pelinovsky (1984) and Pelinovsky and Mazova (1992). A more interesting case occurs when the negative wave (trough) comes near the wall. The nonlinear effects become stronger when the total depth tends to zero. The algebraic solution (5.74) exists only if the trough amplitude is less than $3D/4$; in other words, if the total depth under the trough is greater than $D/4$.

The process of the wave interaction with a vertical wall has been considered for a pulse-like shape of a certain polarity, but this restriction is actually unnecessary. The expression (5.74) can be obtained for an arbitrary function $\eta(T)$, finite or continuous in time, if its shape is sufficiently smooth. The conditions of application of the derived relation between the amplitude of the water oscillations at the wall and the incident wave amplitude are discussed in Pelinovsky et al. (2008). It is shown that the analytical expression (5.74) is valid at least for smooth incident waves if the crest amplitude is less than $3D$ and the trough amplitude is less than $5D/9$. These criteria are obtained from the shallow-water theory, which does not include wave dispersion. Within the framework of the nonlinear-dispersive theory, the height of steady-state waves (cnoidal or solitary waves) is limited as $H = 2A < 0.78D$. According to many laboratory data, where the role of dispersion is important, the wave height is bounded by $0.55D$ (see Massel 1996b). Further, we will use the closed criterion for the normalized significant wave height/depth, $H_s/D < 0.5 \div 0.7$.

The approach applied above is valid for any incident wave that is regular, as well as irregular, due to the wave separation along characteristics. In the latter case, it can be used to analyze distribution functions of the wave field and its spectrum. Unfortunately, it cannot predict the time-lag between the incident wave and water oscillations at the wall, and therefore the function $\eta_w(\tau)$ is not fully determined within the framework of the nonlinear theory. The process is not Gaussian due to nonlinearity, and all the moments cannot be calculated, including the significant height of water oscillations at the wall. On the other hand, the relationship between random wave amplitudes of the incident wave and water oscillations at the wall (see Eq. (5.74)) is explicit and does not include the time-lag. Hence, as soon as the distribution function of the wave amplitude of the incident wave field is known, expression (5.74) can be used to obtain the distribution function of the amplitude of the water oscillations at the wall. The noninertial ("instant") transformation of random processes is described in various books (see Massel 1996a). The exceedance probability function of the water oscillation amplitude at the wall can be determined as follows

$$P_R(R) = P_A(A)|_{A(R)}, \qquad (5.75)$$

where $A(R)$ is the inverse function obtained from Eq. (5.74), which is known explicitly as

$$\frac{A}{D} = \frac{R}{4D} + \frac{1}{2}\left[\sqrt{1+\frac{R}{D}} - 1\right]. \qquad (5.76)$$

For detailed calculations, the exceedance probability function of the incident wave should be specified. Below, the Rayleigh distribution for wave heights is used (indices of distribution functions will be omitted in the following formulas)

$$P(H) = \exp\left(-\frac{H^2}{8\sigma^2}\right) \approx \exp\left(-\frac{2H^2}{H_s^2}\right), \qquad (5.77)$$

where the significant wave height, $H_s \approx 4\sigma$, and σ^2, is the variance of the initial Gaussian field (see formula (2.84) from Chap. 2). In fact, the wave field in shallow water (as well as in deep water) is not Gaussian (see Sect. 5.3), but for the sake of simplicity we will use the assumption of a narrow-band Gaussian process resulting in the Rayleigh distribution for wave heights. For a quasi-monochromatic wave $H = 2A$, the amplitude distribution has the same form as Eq. (5.77). As a result, the exceedance probability functions of the positive (crest) and negative (trough) amplitudes of water oscillations at the vertical wall can be determined explicitly

$$P(R_+) = \exp\left\{-\frac{2}{A_s^2}\left[\frac{R_+}{4} + \frac{1}{2}\left(\sqrt{D+R_+} - D\right)\right]^2\right\}, \qquad (5.78)$$

$$P(R_-) = \exp\left\{-\frac{2}{A_s^2}\left[\frac{R_-}{4} - \frac{1}{2}\left(\sqrt{D-R_-} - D\right)\right]^2\right\}, \qquad (5.79)$$

where both amplitudes (heights of the crests and troughs) have positive values. For the convenience of graphic representation of the distribution functions, the amplitudes of the water oscillations at the wall will hereafter be normalized by $H_s = 2A_s$, taking into account that the wave amplitude on the wall is within the framework of the linear theory twice the amplitude of the incident wave. In this case, any deviation from the Rayleigh distribution characterizes nonlinear effects, and the main parameter here is $\varepsilon = H_s/D$, which is the natural nonlinear parameter of the shallow-water theory.

Fig. 5.22 Exceedance probability function of crest heights of water oscillations at the wall. Numbers on curves denote values of $\varepsilon = H_s/D$ with increment of $\varepsilon = 0.1$

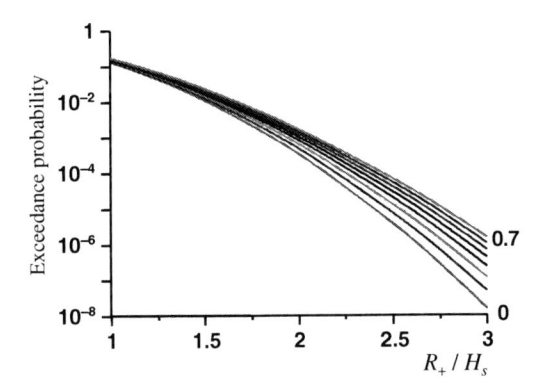

Figure 5.22 displays the exceedance probability function of the crest heights of the water oscillations at the wall for different values of the parameter ε, from 0 (linear case) to 0.7 (large-amplitude waves). As it is expected, weak and moderate water oscillations have almost the same Rayleigh distribution as the incident wave, but their crest heights are twice the incident wave amplitudes (this factor is included in the normalization). For extreme waves, including freak waves (their amplitude exceeds twice and more the significant wave height), the probability of the large crests is increased with an increase of the ratio of the significant wave height to water depth. This means that anomalous high crests should occur in the coastal zone more often than in the open sea, and this effect is related to the nonlinear mechanism of wave transformation in the coastal zone. Such waves may overflow through breakwaters and flood the coasts, causing the accidents described in the literature.

In this way, statistical characteristics of trough amplitudes and wave heights are calculated in Pelinovsky et al. (2008). The probability of occurrence of the deepest troughs near the wall is less than the Rayleigh prediction, and therefore freak waves should often have the shape of crests rather than of troughs. Concerning wave height, it can be concluded that nonlinearity decreases the probability of the highest waves compared with the Rayleigh distribution. It means that the probability of meeting unusual high waves for ships and boats near rocks and breakwaters is less than in the open sea, but the shallow water waves may be significantly steeper due to shoaling effects.

5.5.2 Wave Run-up on a Plane Beach

A similar approach can be applied for the process of long wave run-up on a plane beach, defined by the bottom profile function $D(X) = -\alpha X$ (Fig. 5.23). In this case, the nonlinear shallow-water equations (5.65) can be solved with the use of Riemann invariants

$$I_\pm = u \pm 2\sqrt{g(D+\eta)} + \alpha T \qquad (5.80)$$

Fig. 5.23 Definition sketch for the wave runup problem

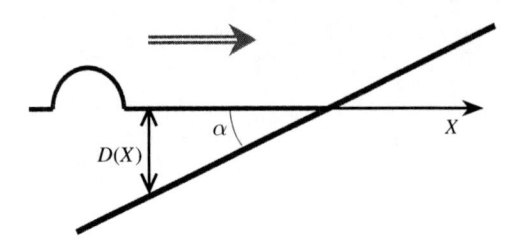

and the Legendre (hodograph) transformation (Carrier and Greenspan 1958). As a result, the long wave run-up process is described by the linear wave equation

$$\frac{\partial^2 \Phi}{\partial \lambda^2} - \frac{\partial^2 \Phi}{\partial \sigma^2} - \frac{1}{\sigma}\frac{\partial \Phi}{\partial \sigma} = 0, \quad (5.81)$$

and all the physical variables can be expressed through the function $\Phi(\lambda,\sigma)$:

$$\eta = \frac{1}{2g}\left(\frac{\partial \Phi}{\partial \lambda} - u^2\right), \quad u = \frac{1}{\sigma}\frac{\partial \Phi}{\partial \sigma}, \quad (5.82)$$

$$T = \frac{1}{\alpha g}\left(\lambda - \frac{1}{\sigma}\frac{\partial \Phi}{\partial \sigma}\right), \quad X = \frac{1}{2\alpha g}\left(\frac{\partial \Phi}{\partial \lambda} - u^2 - \frac{\sigma^2}{2}\right). \quad (5.83)$$

The physical meaning of the variable σ is the total water depth, and $\sigma = 0$ corresponds to the moving shoreline. Various calculations of the wave field and run-up characteristics using the Carrier-Greenspan transformation can be found in Spielfogel (1976), Pedersen and Gjevik (1983), Synolakis (1987), Pelinovsky and Mazova (1992), Tadepalli and Synolakis (1994), Carrier et al. (2003), Tinti and Tonini (2005), Kânoulu and Synolakis (2006), Didenkulova et al. (2006, 2007), and Didenkulova and Pelinovsky (2008).

A surprising result, which follows from the linear equation (5.81), is that the extreme run-up characteristics (run-up and run-down amplitudes, run-up velocities) can be calculated in the framework of the linear shallow-water theory when the incident wave propagates to the beach from the open sea. Particularly, the run-up amplitude of incident sine wave with amplitude A and frequency Ω given on depth D is

$$\frac{R}{A} = \left(\frac{\pi^2 \Omega^2 D}{g\alpha^2}\right)^{1/4}. \quad (5.84)$$

Moreover, the water oscillations on shore are not sinusoidal (see Fig. 5.24). In the figure, cases of various initial amplitudes are shown, expressed through the parameter $Br = R\Omega^2/g\alpha^2$ (condition $Br = 1$ corresponds to wave breaking on shore).

Formulae (5.81), (5.82), (5.83) and (5.84) can be applied to describe the run-up of regular as well as irregular long waves. Due to the implicit character of the Carrier-Greenspan transformation, it is a tricky task to calculate wave characteristics and wave statistics. However, the linear approach may be applied for calculations of the extreme run-up characteristics. Extremes of the Fourier series

5.5 Anomalous High Waves on a Beach

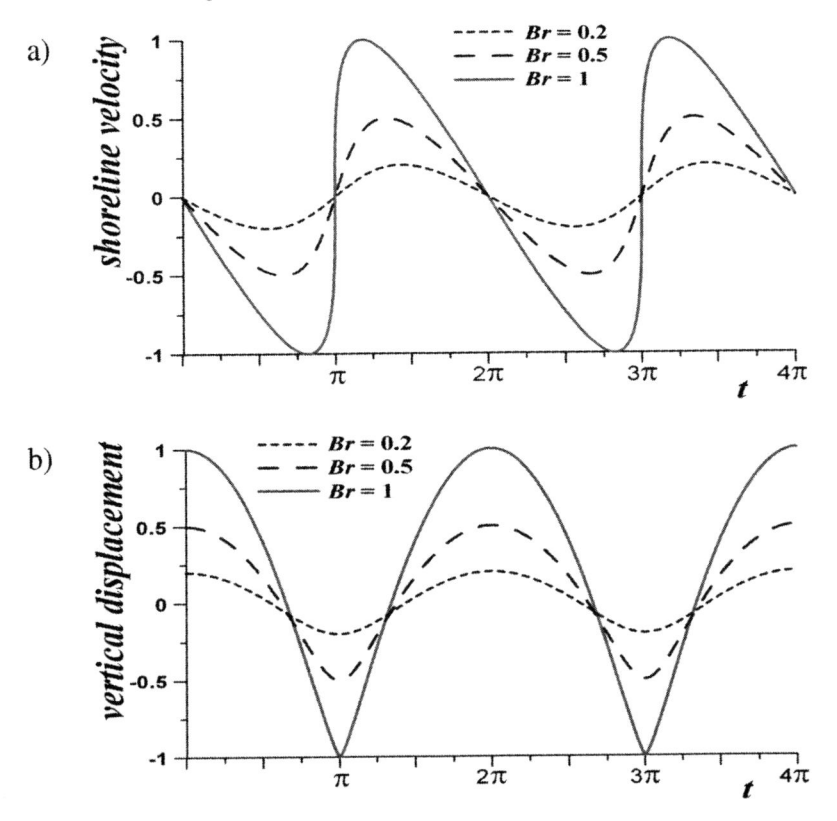

Fig. 5.24 Velocity (**a**) and vertical displacement (**b**) of the moving shoreline

$$\eta(T, X = 0) = \left(\frac{16\pi^2 \Omega^2 D}{g\alpha^2}\right)^{1/4} \sum_{n=1}^{\infty} \sqrt{n} A_n \sin\left[n\Omega(T - \tau) + \frac{\pi}{4}\right], \quad (5.85)$$

should be obtained for this purpose (Didenkulova et al. 2007, Didenkulova and Pelinovsky 2008). In Eq. (5.85), A_n denotes the spectral amplitudes, Ω is the basic frequency of the incident wave, and τ is the travel time to the coast.

It should be emphasized that series (5.85) can be used when calculating positive and negative run-up amplitudes, but not the moments and distribution functions of the water displacement onshore. Detailed calculations of the distribution functions of the run-up amplitudes are given in(Sergeeva and Didenkulova (2005). For the narrow-band incident wave field, the distribution functions of the run-up characteristics are described by the Rayleigh distribution, as is expected owing to the linearity of the expressions for extreme characteristics. When the spectrum of the incident wave is wider, the distribution functions differ from the Rayleigh law; the mean value of the run-up amplitude changes as well.

The wave field in shallow water involves many coherent wave components. A way to represent such a field as a random set of solitary waves is very popular (see

Brocchini and Gentile 2001). The run-up of a solitary wave on a plane beach is well studied (Synolakis 1987), and the run-up amplitude, R, can be expressed through the soliton amplitude, A, as

$$\frac{R}{D} = 2.8312 \frac{1}{\sqrt{\alpha}} \left(\frac{A}{D}\right)^{5/4}. \tag{5.86}$$

In fact, this formula can be derived from Eq. (5.85) by taking into account the relation between the soliton amplitude and the duration. When the wave field contains random separated solitons, the runup of each individual soliton represents an independent random process and the distribution function of run-up amplitude can be found analytically when the distribution function of the soliton amplitudes is known. Assuming for the sake of simplicity that the Rayleigh distribution for the soliton amplitude, and using (5.86), the exceedance probability of run-up amplitude is

$$P(R) = \exp\left[-0.378\alpha^{4/5} \frac{(R/D)^{8/5}}{(A/D)^2}\right]. \tag{5.87}$$

The tail of the distribution (5.87) decays slower than that of the Rayleigh distribution. Therefore, the probability of large wave occurrence on coasts is high. More detailed computations of statistical run-up characteristics of the wave field represented by a soliton ensemble are performed in Brocchini and Gentile (2001).

So, the wave run-up on a vertical wall or plane beach leads to an increase of the probability of large-amplitude waves. Thus, a way to reduce possible rogue wave damage should be to include proper coastal protection.

List of Notations

A	wave amplitude
A_s	significant wave amplitude
$b(X,Y)$	distance between neighbouring rays
C	long-wave speed
D	water depth
g	acceleration due to gravity
H	wave height
H_s	significant wave height
$I_\pm$	Riemann invariants
k	dimensionless wavenumber
K	wavenumber
l	coordinate along the ray
N_s	soliton number
P	probability distribution function
R	runup amplitude
s	temporal variable
S	non-symmetric wave spectrum

t	dimensionless time
T	time
$\mathbf{u}(X,Y,T)$	depth-averaged velocity
$\mathbf{U}=(U,V)$	fluid velocity in the horizontal plane
Ur	Ursell parameter
V	velocity of the soliton
(x,y)	dimensionless coordinates in the horizontal plane
(X,Y,Z)	coordinates
W	vertical fluid velocity
ε	nonlinear parameter
$\phi(X,Y,Z,T)$	velocity potential
γ	skewness
$\eta(X,Y,T)$	surface elevation
κ	kurtosis
$\tilde{\kappa}$	normalized kurtosis
λ	wavelength
σ	depth variable in the hodograph transformation
σ	standard deviation, σ^2 is the variance
Ω	cyclic wave frequency
$\zeta(x,y,t)$	dimensionless surface displacement
∇	gradient operator in the horizontal plane

References

Agnon Y, Madsen PA, Schaffer HA (1999) A new approach to high order Boussinesq models. J Fluid Mech 399:319–333

Brocchini M, Gentile R (2001) Modelling the run-up of significant wave groups. Cont Shelf Res 21:1533–1550

Carrier GF, Greenspan HP (1958) Water waves of finite amplitude on a sloping beach. J Fluid Mech 4:97–109

Carrier GF, Wu TT, Yeh H (2003) Tsunami run-up and draw-down on a plane beach. J Fluid Mech 475:79–99

Chen Q, Kirby JT, Dalrymple RA et al (2000) Boussinesq modeling of wave transformation, breaking, and run-up. J Waterway Port Coast Ocean Eng 126:48–56

Didenkulova I, Pelinovsky E (2008) Runup of solitary waves of various shapes on a beach. Oceanology 48:5–10

Didenkulova I, Pelinovsky E, Soomere T, Zahibo N (2007) Runup of nonlinear asymmetric waves on a plane beach. In: Kundu A (ed) Tsunami and Nonlinear Waves, pp 173–188. Springer

Didenkulova II, Zahibo N, Kurkin AA, Levin BV, Pelinovsky EN, Soomere T (2006) Runup of nonlinearly deformed waves on a coast. Dokl Earth Sci 411:1241–1243

Dingemans MW (1996) Water waves propagation over uneven bottom. Vol 2. World Sci, Singapore

Drazin PG, Johnson RS (1989) Solitons: an Introduction. Cambridge University Press, Cambridge

Dysthe KB, Trulsen K, Krogstad HE, Socquet-Juglard H (2003) Evolution of a narrow-band spectrum of random surface gravity waves. J Fluid Mech 478:1–10

Engelbrecht JK, Fridman VE, Pelinovski EN (1988) Nonlinear evolution equations. Longman, London

Fornberg B (1998) A Practical guide to pseudospectral methods. Cambridge University Press, Cambridge

Funakoshi M (1980) Reflection of obliquely incident large-amplitude solitary waves. J Phys Soc Japan 49:2371–2379
Gardner CS, Green JM, Kruskal MD, Miura RM (1967) Method for solving the Korteweg – de Vries equation. Phys Rev Lett 19:1095–1097
Green AE, Naghdi PM (1976) A derivation of equations for wave propagation in water of variable depth. J Fluid Mech 78:237–246
Grimshaw R, Pelinovsky D, Pelinovsky E, Talipova T (2001) Wave group dynamics in weakly nonlinear long-wave models. Phys D 159:35–57
Groesen E, Westhuis JH (2002) Modeling and simulation of surface water waves. Math Comput Simul 59:341–360
Janssen PAEM (2003) Nonlinear four-wave interactions and freak waves. J Phys Oceanogr 33:863–884
Kânoulu U, Synolakis C (2006) Initial value problem solution of nonlinear shallow water-wave equations. Phys Rev Lett 97:148501
Kharif C, Pelinovsky E, Talipova T (2000) Formation de vagues géantes en eau peu profonde. Comptes Rendus de l'Académie des Sciences 328, série Iib:801–807
Kim JW, Bai KJ, Ertekin RC, Webster WC (2003) A strongly-nonlinear model for water waves in water of variable depth – the irrotational Green-Naghdi model. J Offshore Mech Arctic Eng 125:25–32
Kit E, Shemer L, Pelinovsky E, Talipova T, Eitan O, Jiao H (2000) Nonlinear wave group evolution in shallow water. J Waterway Port Cost Ocean Eng 126:221–228
Kokorina A, Pelinovsky E (2005) Numerical simulation of the random waves in shallow water using experimental data. In: Proc 3rd Int Conf APAC (Korea), 1334–1345
Madsen PA, Schaffer HA (1998) Higher-order Boussinesq-type equations for surface gravity waves: derivation and analysis. Phil Trans Roy Soc Lond A 356:3123–3184
Madsen PA, Bingham HB, Liu H (2002) A new Boussinesq method for fully nonlinear waves from shallow to deep water. J Fluid Mech 462:1–30
Madsen PA, Bingham HB, Schaffer HA (2003) Boussinesq-type formulations for fully nonlinear and extremely dispersive water waves: derivation and analysis. Proc Roy Soc Lond A 459:1075–1104
Mase H, Memita T, Yuhi M, Kitano T (2002) Stem waves along the vertical wall due to random wave incidence. Coast Eng 44:339–350
Massel SR (1996a) Ocean surface waves: their physics and prediction. World Scientific Publishing Co Pte Ltd, Singapore
Massel SR (1996b) On the largest wave height in water of constant depth. Ocean Eng 23:553–573
Matveev VB (2002) Positons: slowly decreasing analogues of solitons. Theor Math Phys 131:483–497
Melville WK (1980) On the Mach reflection of solitary waves. J Fluid Mech 98:285–297
Miles JW (1977a) Obliquely interacting solitary waves. J Fluid Mech 79:157–169
Miles JW (1977b) Resonantly interacting solitary waves. J Fluid Mech 79:171–179
Mirchina N, Pelinovsky E (1984) Increase in the amplitude of a long wave near a vertical wall. Izv Atmos Ocean Phys 20:252–253
Murray AC (1978) Solutions of the Korteweg – de Vries equation from irregular data. Duke Math J 45:149–181
Novikov S, Manakov SV, Pitaevskii LP, Zakharov VE (1984) Theory of Solitons: the Inverse Scattering Method. Consult Bureau, New York
Onkuma K, Wadati M (1983) The Kadomtsev – Petviashvili equation, the trace method and the soliton resonance. J Phys Soc Japan 52:749–760
Onorato M, Osborne AR, Serio M, Bertone S (2001) Freak waves in random oceanic sea states. Phys Rev Lett 86:5831–5834
Onorato M, Osborne AR, Serio M, Cavaleri L (2005) Modulational instability and non-Gaussian statistics in experimental random water-wave trains. Phys Fluids 17:078101-1–4
Osborne AR (1995) Solitons in the periodic Korteweg – de Vries equation, the Θ-function representation, and the analysis of nonlinear, stochastic wave trains. Phys Rev E 52:1105–1122

Osborne AR, Serio M, Bergamasco L, Cavaleri L (1998) Solitons, cnoidal waves and nonlinear interactions in shallow-water ocean surface waves. Phys D 123:64–81

Ostrovsky LA, Pelinovsky EN (1975) Refraction of nonlinear ocean waves in a beach zone. Izv Atmos Ocean Phys 11:37–41

Pedersen G, Gjevik B (1983) Runup of solitary waves. J Fluid Mech 142:283–299

Pelinovsky E, Talipova T, Kharif C (2000) Nonlinear dispersive mechanism of the freak wave formation in shallow water. Phys D 147:83–94

Pelinovsky EN (1982) Nonlinear dynamics of tsunami waves. IAP RAS Press, Nizhny Novgorod (In Russian)

Pelinovsky E, Mazova R (1992) Exact analytical solutions of nonlinear problems of tsunami wave run-up on slopes with different profiles. Nat Hazards 6:227–249

Pelinovsky E, Kharif C, Talipova T (2008) Large-amplitude long wave interaction with a vertical wall. Eur J Mech B/Fluids 27:409–418

Pelinovsky E, Sergeeva (Kokorina) A (2006) Numerical modeling of the KdV random wave field. Eur J Mech B/Fluids 25:425–434

Pelinovsky EN (1996) Hydrodynamics of tsunami waves. IAP RAS Press, N Novgorod (In Russian)

Peregrine DH (1967) Long waves on a beach. J Fluid Mech 27:815–827

Peregrine DH (1972) Equaions for water waves and approximations behind them. In: Meyer R (ed) Waves on Beaches. New York Academic Press, New York, pp 95–122

Peterson P, Soomere T, Engelbrecht J, van Groesen E (2003) Interaction solitons as a possible model for extreme waves in shallow water. Nonlin Proc Geophys 10:503–510

Porubov AV, Tsuji H, Lavrenov IV, Oikawa M (2005) Formation of the rogue wave due to nonlinear two-dimensional waves interaction. Wave Motion 42:202–210

Salupere A, Maugin GA, Engelbrecht J, Kalda J (1996) On the KdV soliton formation and discrete spectral analysis. Wave Motion 123:49–66

Salupere A, Peterson P, Engelbrecht J (2002) Long-time behaviour of soliton ensembles. Part 1 – Emergence of ensembles. Chaos Solitons Fractals 14:1413–1424

Salupere A, Peterson P, Engelbrecht J (2003a) Long-time behaviour of soliton ensembles. Part 2 – Periodical patterns of trajectorism. Chaos Solitons Fractals 15:29–40

Salupere A, Peterson P, Engelbrecht J (2003b) Long-time behavior of soliton ensembles. Math Comput Simul 62:137–147

Satsuma J (1976) N-soliton solution of the two-dimensional Korteweg–de Vries equation. J Phys Soc Japan 40:286–290

Sergeeva AV, Didenkulova II (2005) Runup of irregular waves on a plane beach. Proc Ac Eng Sc Russ Fed 14:98–105 (In Russian)

Soomere T, Engelbrecht J (2005) Extreme evaluation and slopes of interacting solitons in shallow water. Wave Motion 41:179–192

Spielfogel LO (1976) Runup of single waves on a sloping beach. J Fluid Mech 74:685–694

Synolakis CE (1987) The runup of solitary waves. J Fluid Mech 185:523–545

Tadepalli S, Synolakis CE (1994) The Runup of N-waves. Proc Roy Soc London A 445:99–112

Talipova T, Kharif C, Giovanangeli J-P (2008) Modelling of rogue wave shapes in shallow water. In: Pelinovsky E, Kharif C (eds) Extreme Ocean Waves, Springer 71–81

Tanaka M (2001) A method of studying of nonlinear random field. Fluid Dyn Res 28:41–60

Tinti S, Tonini R (2005) Analytical evolution of tsunamis induced by near-shore earthquakes on a constant-slope ocean. J Fluid Mech 535:33–64

Toffoli A, Onorato M, Osborne AR, Monbaliu J (2006) Non-gaussian properties of surface elevation in crossing sea states in shallow water. In: Proc Int Coast Eng (ICCE06), 782–790

Wei G, Kirby JT, Grilli ST, Subramanya R (1995) A fully nonlinear Boussinesq model for free surface waves. Part 1: Highly nonlinear unsteady waves. J Fluid Mech 294:71–92

Zabusky NJ, Kruskal MD (1965) Interaction of solutions in a collisionless plasma and recurrence of initial states. Phys Rev Lett 15:240–243

Zakharov VE (1971) Kinetic equation for solitons. Sov J Exp Theor Phys 60:993–1000

Zheleznyak MI, Pelinovsky EN (1985) Physical and mathematical models of the tsunami climbing a beach. In: Tsunami Climbing a Beach. IAP AS Press, Gorky, 8–34

Chapter 6
Conclusion

In this book, we have collected some of the most reliable in-situ observations of rogue waves by people, and instrumental registrations of abnormal waves during long-term wave recordings. Until now, the observations have been sufficient to prove the existence of this extreme wave phenomenon, and to show some of its most pronouncing features, such as the sudden huge wave growth that often could not be foreseen by the usual experience of navigating. Additionally, the existing natural observations do not allow us to create a definite theory of these waves.

Different physical mechanisms have been suggested as possible explanations of this phenomenon. These mechanisms are presented and described in the book in detail. They are based on different kinds of wave focusing—namely, spatial and spatio-temporal wave focusing due to inhomogeneities existing in basins and water wave dispersion, nonlinear wave dynamics, and nonlinear instabilities. The occurrence of rogue waves due to the influence of current and wind action has been considered as well. While simplified mathematical models may help us to qualitatively understand the main physical effects, they become questionable when huge waves occur because of strong nonlinear effects. Hence, the use of the exact equations is inevitable in correctly describing the dynamics of this strongly nonlinear event. The use of fully nonlinear models is now relevant in studying extreme wave events thanks to both the recent improvement of numerical methods, and the development of computer performance.

Some physical effects, resulting in strong wave enhancement, may be relatively easily reproduced in the laboratory. We described various laboratory experiments on rogue waves due to dispersive wave focusing, with and without wind and current influences. These results are compared with theoretical achievements and fully nonlinear numerical simulations. Other mechanisms (such as nonlinear wave instabilities and directional effects) could be reproduced in the laboratory with much more effort. Therefore, there are not many, and the majority of experiments are performed in numerical wave tanks. Use of efficient numerical codes also looks promising when investigating statistics of high waves.

In spite of doubtless progress in the understanding, description, and forecasting of steep waves, many questions still remain. The improvement of existing theories turned out to be insufficiently supported by natural observations. This lack of data makes it difficult to check the validity of the developed theories, and to create statistical models based on in-situ data. One of the main challenges is to find which of the statistical parameters are the most relevant to define the occurrence probability of freak waves correctly. For instance, it is well accepted that kurtosis is a good indicator of rogue wave generation in 2D sea states, while it is far from a good indicator for the case of 3D water waves. To answer this problem, it is necessary to go beyond approximate equations even if approximate models may give satisfactory results for specific conditions. Hence, the need for fast numeric codes based on the exact equations for investigating statistical features of extreme wave events in 3D geometries becomes urgent.

Direct observation of extreme water waves from satellites should be very useful, even though this task is not completely resolved yet. From a theoretical viewpoint, it should be interesting to couple hydrodynamic and electromagnetic codes to determine which signature a rogue wave leaves in the backscattered electromagnetic signal by the sea surface.

On the other hand, the interest created by rogue waves in the ocean has stimulated investigations of "rogue waves" in other situations. First, let us mention other kinds of rogue waves in geophysics. In Chap. 1, coastal freak waves were discussed. Although these waves have attributes similar to rogue waves in the open sea—i.e., size and sudden appearance—they obviously require different theoretical models and may result from different physical mechanisms. The effects of dispersive focusing can arise in tsunami-wave fields of seismic or volcanic origin, when multiple shocks or explosions occur (Mirchina and Pelinovsky 2001). Trapped edge waves were supposed responsible for freak coastal events in Kurkin and Pelinovsky (2002). Trapped waves may undergo dispersive and modulational focusing (described in Chaps. 3 and 4). Anomalously high internal waves in the stratified sea, and planetary Rossby waves, may potentially happen due to dispersion and modulational instability as well (Kurkin and Pelinovsky 2004). Furthermore, since the Korteweg-de Vries (KdV) equation and the nonlinear Schrödinger equation are universal equations that apply to many other physical fields, similar mechanisms resulting in rogue waves are expected to occur in these fields. We mention the experimental and numerical investigations in nonlinear optics, where optical rogue waves can occur in fibers (Solli et al. 2007, Yeom and Eggleton 2007). Huge waves have been shown to occur in plasmas as well (Ruderman et al. 2008). These nonlinear physical processes are similar to those generating rogue waves from modulational instability.

References

Kurkin AA, Pelinovsky EN (2002) Focusing of edge waves above a sloping beach. Eur J Mech B / Fluids 21:561–577
Kurkin AA, Pelinovsky EN (2004) Freak waves: facts, theory and modelling. NNSTU Press, N Novgorod (In Russian)
Mirchina N, Pelinovsky E (2001) Dispersive intensification of tsunami waves. In: Proc Int Tsunami Conf, Seattle 7-9 August 2001, pp 789–794
Ruderman MS, Talipova T, Pelinovsky E (2008) Dynamics of modulationally unstable ion-acoustic wave packets in plasmas with negative ions. J Plasma Phys. doi: 10.1017/S0022377808007150 vol. 74, No 5, 639–656
Solli DR, Ropers C, Koonath P, Jalali B (2007) Optical rogue waves. Nature 450:1054–1057
Yeom D-II Eggleton BJ (2007) Rogue waves surface in light. Nature 450:953–954

Appendix A
Discretisation of the Boundary Integral Equation for the Potential

In this appendix, we show how Eqs. (4.100) and (4.101) may be transformed into a linear system of algebraic equations.

The boundaries $\partial \Omega_{FS}$ and $\partial \Omega_{SB}$ are discretized into N_F and N_B panels, s_j, respectively. Hence, the discretized version of Eqs. (4.100) and (4.101) is given by the following system

$$\theta_i \varphi_i - \sum_{j=1}^{N_F} \int_{s_j} \varphi_j \frac{\partial G}{\partial n}(i,j)\,ds + \sum_{j=1}^{N_B} \int_{s_j} \frac{\partial \varphi_j}{\partial n} G(i,j)\,ds$$
$$= \sum_{j=1}^{N_F} \int_{s_j} \frac{\partial \varphi_j}{\partial n} G(i,j)\,ds + \sum_{j=1}^{N_B} \int_{s_j} \varphi_j \frac{\partial G}{\partial n}(i,j)\,ds, \qquad (A.1)$$

where $1 < i < N_F + 1$ corresponds to point i of the free surface, and

$$\sum_{j=1}^{N_F} \int_{s_j} \varphi_j \frac{\partial G}{\partial n}(i,j)\,ds + \sum_{j=1}^{N_B} \int_{s_j} \frac{\partial \varphi_j}{\partial n} G(i,j)\,ds$$
$$= \theta_i \varphi_i + \sum_{j=1}^{N_F} \int_{s_j} \frac{\partial \varphi_j}{\partial n} G(i,j)\,ds - \sum_{j=1}^{N_B} \int_{s_j} \varphi_j \frac{\partial G}{\partial n}(i,j)\,ds, \qquad (A.2)$$

where $1 < i < N_B + 1$ corresponds to point i of the solid boundary.

For a 2D motion, the Green function is written as follows,

$$G(P,Q) = \ln\left(\left|\overrightarrow{PQ}\right|\right). \qquad (A.3)$$

It is interesting to note that this function introduces only the distance separating the points P and Q. Hence, by introducing the local coordinates of the panels (ξ, ϑ), this function can be rewritten as follows:

$$G(P,Q) = \ln\left(\sqrt{\xi^2 + \vartheta^2}\right). \qquad (A.4)$$

Hence, the normal derivative can be obtained:

$$\frac{\partial G}{\partial n}(P,Q) = \frac{\vartheta}{\xi^2 + \vartheta^2}. \qquad (A.5)$$

When assuming a linear variation of φ and $\partial \varphi / \partial n$ along the panels, Eq. (A.1) becomes

$$\theta_i \varphi_i - \sum_{j=1}^{N_F} \left\{ \varphi_{j+1} \frac{I_4 - \xi_j I_2}{\xi_{j+1} - \xi_j} + \varphi_j \frac{I_2 \xi_{j+1} - I_4}{\xi_{j+1} - \xi_j} \right\} + \sum_{j=1}^{N_B} \left\{ \psi_{j+1} \frac{I_3 - \xi_j I_1}{\xi_{j+1} - \xi_j} + \psi_j \frac{I_1 \xi_{j+1} - I_3}{\xi_{j+1} - \xi_j} \right\}$$

$$= \sum_{j=1}^{N_F} \left\{ \psi_{j+1} \frac{I_3 - \xi_j I_1}{\xi_{j+1} - \xi_j} + \psi_j \frac{I_1 \xi_{j+1} - I_3}{\xi_{j+1} - \xi_j} \right\} + \sum_{j=1}^{N_B} \left\{ \varphi_{j+1} \frac{I_4 - \xi_j I_2}{\xi_{j+1} - \xi_j} + \varphi_j \frac{I_2 \xi_{j+1} - I_4}{\xi_{j+1} - \xi_j} \right\}$$

$$(A.6)$$

and Eq. (A.2)

$$\sum_{j=1}^{N_F} \left\{ \varphi_{j+1} \frac{I_4 - \xi_j I_2}{\xi_{j+1} - \xi_j} + \varphi_j \frac{I_2 \xi_{j+1} - I_4}{\xi_{j+1} - \xi_j} \right\} + \sum_{j=1}^{N_B} \left\{ \psi_{j+1} \frac{I_3 - \xi_j I_1}{\xi_{j+1} - \xi_j} + \psi_j \frac{I_1 \xi_{j+1} - I_3}{\xi_{j+1} - \xi_j} \right\}$$

$$= \theta_i \varphi_i + \sum_{j=1}^{N_F} \left\{ \psi_{j+1} \frac{I_3 - \xi_j I_1}{\xi_{j+1} - \xi_j} + \psi_j \frac{I_1 \xi_{j+1} - I_3}{\xi_{j+1} - \xi_j} \right\}$$

$$- \sum_{j=1}^{N_B} \left\{ \varphi_{j+1} \frac{I_4 - \xi_j I_2}{\xi_{j+1} - \xi_j} + \varphi_j \frac{I_2 \xi_{j+1} - I_4}{\xi_{j+1} - \xi_j} \right\}, \qquad (A.7)$$

where ψ corresponds to the normal derivative of the potential, $\psi = \partial \varphi / \partial n$, and

$$I_1 = \int_{\xi_1}^{\xi_2} \ln\left(\sqrt{x^2 + \vartheta^2}\right) dx = \frac{1}{2} \xi_2 \ln\left(\xi_2^2 + \vartheta^2\right) - \xi_2$$

$$+ \vartheta \arctan\left(\frac{\xi_2}{\vartheta}\right) - \frac{1}{2} \xi_1 \ln\left(\xi_1^2 + \vartheta^2\right) + \xi_1 - \vartheta \arctan\left(\frac{\xi_1}{\vartheta}\right),$$

$$I_2 = \int_{\xi_1}^{\xi_2} \frac{\vartheta}{x^2 + \vartheta^2} dx = \arctan\left(\frac{\xi_2}{\vartheta}\right) - \arctan\left(\frac{\xi_1}{\vartheta}\right),$$

$$I_3 = \int_{\xi_1}^{\xi_2} x \ln\left(\sqrt{x^2 + \vartheta^2}\right) dx = \frac{1}{4} \xi_2^2 \ln\left(\xi_2^2 + \vartheta^2\right) + \frac{1}{4} \vartheta^2 \ln\left(\xi_2^2 + \vartheta^2\right)$$

$$- \frac{1}{4} \xi_2^2 - \frac{1}{4} \xi_1^2 \ln\left(\xi_1^2 + \vartheta^2\right) - \frac{1}{4} \vartheta^2 \ln\left(\xi_1^2 + \vartheta^2\right) + \frac{1}{4} \xi_1^2,$$

$$I_4 = \int_{\xi_1}^{\xi_2} \frac{x \vartheta}{x^2 + \vartheta^2} dx = \frac{1}{2} \vartheta \ln\left(\xi_2^2 + \vartheta^2\right) - \frac{1}{2} \vartheta \ln\left(\xi_1^2 + \vartheta^2\right).$$

Index

Abnormality index, *AI*, definition, 7
Amplication factor, 8, 22, 23, 68, 80, 81, 115, 116, 129, 131–134, 145

Benjamin–Feir index, 140–143, 147, 158, 162, 163
Benjamin–Feir instability, *See* Modulational (Benjamin–Feir) instability, 8
Blocking point, 82–85
Bound waves, 6, 47, 48, 55, 102, 116, 161
Boussinesq equation, 175
Breathers, 111–116, 121, 135, 149

Caustics, 58, 66–69, 71, 72, 82, 83, 85, 86, 176, 198
Chaos, 43, 115–117, 140, 184
Cnoidal waves, 57, 173, 183, 184, 188, 198, 201
Crescent-shaped instability, 95, 149–151

Davey-Stweartson equation, DS, 98–101, 103, 149
Dispersive parameter, 52, 57, 97, 174, 192
Downshifting, 123, 126, 135, 140, 189, 190
Dysthe equation, 92, 135, 140, 141, 152, 190

Eikonal equation, 176
Energy balance equation, 65, 71, 82, 155

Fermi-Pasta-Ulam recurrence, 94
Fetch, 6, 7, 28, 29, 52, 53, 78

Gaussian distribution, 46, 73–75, 79, 103, 109, 185, 190, 191
Green and Naghdi equation, 175

Hole in the sea, v, 17, 22, 23, 75, 147, 179, 181

Instability criterion, 93, 100–105, 141, 143, 147, 154, 158

Inverse Scattering Transform, 91, 106, 159, 177

Jeffreys' sheltering mechanism, 81, 127, 130–133
JONSWAP spectrum, 53, 140, 146, 163

Kadomtsev–Petviashvili equation, 177, 193, 197
Kinematic equations, 54, 64, 65, 69
Korteweg – de Vries equation, KdV, 106, 160, 175–179, 181, 183–186, 196, 212
Kurtosis, 45, 46, 141, 163, 188–191, 197, 212

Lifetime of steep wave event, 18, 22, 132, 197
Long-crested waves, 140, 146, 147, 153, 163

Mach stem, 195–198
Miles' wave amplification mechanism, 79, 80, 130
Mixed sea conditions, 48, 147
Modulational (Benjamin–Feir), 143
Modulational (Benjamin–Feir) instability, vi, 57, 58, 91, 92, 94, 95, 100, 102–105, 111, 116, 117, 121, 122, 127, 134, 141, 146, 147, 149–151, 154, 158–160, 177, 212

Nonlinear parameter, 39, 57, 97, 108, 118, 140, 142, 143, 160, 178, 192, 196, 202, 203
Nonlinear Schrödinger equation, NLS, 92, 95, 99, 100, 103–108, 110, 111, 113, 114, 116, 121, 135, 137–139, 142, 147–150, 152, 154, 158–161

Padé approximation, 175
Peakedness, 45, 53, 146, 163, 188
Peregrine system, 175
Pierson-Moskowitz spectrum, 53
Probability exceedance, 47, 202, 203, 206

Pyramidal waves, 19, 23, 147, 154

Ray theory, 63, 66, 67, 71, 81, 85, 176
Rayleigh distribution, 46–48, 162, 190, 191, 202, 203, 205, 206
Rayleigh–Jeans distribution, 189
Recurrence phenomenon, 94, 116, 117, 123, 125, 135, 149–153, 184
Riemann wave, 70, 199
Rogue waves in other physics, vi, 212

Short-crested waves, 102, 140, 152–154, 163
Significant wave height, definition, 7
Skewness, 45, 46, 163, 188, 190, 191, 197
Solitary waves, vi, 18, 57, 58, 85, 106–116, 135, 139, 142, 143, 147, 158–161, 177–179, 181, 184, 185, 192–197, 206
Solitonic turbulence, 116, 122, 184
Spectral instability, 91, 103
Spectral moments, definition, 51
Statistical moments, definition, 45

Stokes waves, 47, 57, 92, 93, 117, 122–125, 128, 150, 151, 158
Synthetic Aperture Radar (SAR), 3, 11, 26, 161

Unstable modes, 116, 159, 160
Ursell parameter, 142, 178, 179, 184, 185, 188–191

Variance, 45, 46, 103, 141, 185, 188, 202

Walls of water, v, 8, 19, 23, 147
Wave action balance equation, 82
Wave age, 6, 43, 52
Wave height, definition, 1
Wave steepness, definition, 7

Zakharov equation, 55, 92, 103, 116, 119, 140, 151, 154
Zheleznyak and Pelinovsky equation, 175